Young screech owls

AT HOME IN THE WILD

The Story of Animal Habitat

Frances and Dorothy Wood

Illustrated with photographs

DODD, MEAD & COMPANY

NEW YORK

1   2   3   4   5   6   7   8   9   10

Library of Congress Cataloging in Publication Data

Wood, Frances Elizabeth.
   At home in the wild.

   Bibliography: p.
   Includes index.
   SUMMARY: Discusses measures taken to help endangered
species adapt to the changing environment.
   1. Zoology—Ecology—Juvenile literature.
2. Rare animals—Juvenile literature.   [1. Rare
animals.   2. Wildlife conservation]   I. Wood,
Florence Dorothy, joint author.   II. Title.
QH541.14.W66   591.9'73   76-53619
ISBN 0-396-07429-4

ACKNOWLEDGMENTS

For invaluable help in assembling information and for critically reading portions of the manuscript, the authors are grateful to:

Dr. Ray C. Erickson, Assistant Director for Endangered Wildlife Research, Patuxent Wildlife Research Center; Ralph T. Heath, Jr., Founder and President, Suncoast Seabird Sanctuary, Inc.; Mrs. Christine Stevens, President, Animal Welfare Institute; Russell R. Hoffman, Manager, Grays Lake National Wildlife Refuge, U.S. Fish and Wildlife Service; and Wilford O. Nelson, Director, Region II, U.S. Fish and Wildlife Service.

PICTURE CREDITS: California Department of Fish and Game, page 122. Florida News Bureau, Department of Commerce, 12, 14, 15, 24, 28, 35, 37, 102, 107, 111, 147, 170 (left, top and bottom). Michigan Department of Conservation, 136. Montana Highway Commission, 68, 74. National Park Service, 164 (bottom); Isle Royale National Park, William Dunmire, 49; R. Linn, 131. New Mexico Department of Fish and Game, 104. Peter Pritchard, 170 (right). Suncoast Seabird Sanctuary, Stan Ashbrook, 2, 38, 91, 95; Rick Gilson, 92; Jay Morris, 10, 41, 93, 96. U.S. Fish and Wildlife Service, 81, 121; Charles L. Cadieux, 47; Jack F. Dermid, 17; J. S. Dixon, 63; Andrew H. DuPre, 20; Mildred Adaam Fenton, 119; Luther C. Goldman, 19, 80, 84, 99, 101, 113, 168, 177; S. A. Grimes, 40; E. P. Haddon, 46; Henry Harmon, 184 (top); B. M. Hazeltine, 134, 137; C. J. Henry, 27; Robert D. Jones, Jr., 161, 184 (bottom); W. H. Julian, 100; E. R. Kalmbach, 29, 64; W. F. Kubichek, 85; Erwin McIntosh, 58; Miami Seaquarium, 156; Gale Monson, 130; O. J. Murie, 174; National Park Service, 175; Don Reilly, 105; W. M. Rush, 182; V. B. Scheffer, 164 (top); Rex Gary Schmidt, 144; LeRoy W. Sowl, 165; Hal Swiggett, 180; U.S. Coast Guard, 183. U.S. Forest Service, 57, 65. Wyoming Travel Commission, 75.

WHERE DO ANIMALS LIVE?

What kinds of places do animals live in? What kinds of places make homes for them? Think of the places near your own home. Your yard. A vacant lot. A park across the street. A meadow or farm. A woodland. A deep forest. A hillside. A rocky mountainside or a high mountain peak. A sandy seashore.

Think of water places, too. A swamp or marsh. A creek or big river. A pond or lake. A bay. An ocean.

Animals live in all of these different kinds of places. Different kinds of animals live in each place.

We say that an animal is "fitted" for living in its special kind of place. The parts of its body are fitted to that place, and usually they make the animal different from those that live in other places. Its habits are different, too.

Sometimes an animal can live in more than one kind of place. A raccoon, for example, can live in marshy places, or in a meadow, or in a woodland. But many times an animal is so completely fitted to its special kind of place that it will live in no other. Then it demands that special kind, and if the place is destroyed, the animal dies out. Kirtland's warbler, for example, will build its nest in young jack pines, and nowhere else. If the pines go, the warblers go, too.

Nine times out of ten, if an animal becomes extinct, or if it

7

becomes "endangered" or threatened with extinction, something has happened to the animal's home, and it can no longer live there. So in man's efforts to save the endangered animals, he looks first at the homes. He collects facts to show just what kind of home the animal prefers, and how it lives there.

Sometimes man's killing of animals will endanger a certain kind. Perhaps the only change is that man has come, with a gun in his hand. But usually there are other changes—cutting a forest, draining a swamp, plowing a plains area.

So, trying to save the animals, man looks at these places where there have been changes, and tries to collect facts about the animals in them. He studies animals at home.

CONTENTS

Young brown pelican

Chapter 1

HOW PELICANS LIVE

We pulled the car to the side of the road and watched them —fourteen pelicans flying in a straight line over the water. We were on the approach to the Sunshine Skyway in Florida; the approach is a highway that goes from one small island to the next by causeway, separating Tampa Bay from the Gulf of Mexico. In these surroundings, there was nothing unusual about seeing fourteen pelicans in the air at once, flying in a V or a straight line formation.

But what the pelicans were doing was unusual. They would soar, then flap along, in unison. Then, all together, they would pivot in the air, their heads going down, their wings pulling in, and come hurtling downward with a big splash into the water.

This is the technique that pelicans use for fishing, but we didn't see one of them catch a fish. Rather, what they seemed to be doing was a water-and-air ballet, performed for the fun of it. After the big splash, they would sit on the water for a moment, then rise into the air and flap and soar toward another dive, every one in time with all the others.

This may not be a unique happening—but we saw it years ago, and we have never come across anyone else who has seen it, or heard of anyone else seeing it.

11

Pelicans flying together

DECREASING NUMBERS

Today it is a safe bet that no one sees the "ballet" often, because it is seldom that we see more than two or three pelicans in the air together at a time. The brown pelican has made the endangered list of American wildlife, chiefly because of what has happened to it in other states than Florida. In Louisiana, where it is the state bird, it was completely wiped out. In Texas there are only a few pairs. In California the once large population has been reduced to a small fraction of yesterday's number. The same story holds for Mississippi, Alabama, and South Carolina.

Everywhere but in Florida, the pelican seemed headed for quick extinction. Everywhere that the water was polluted with DDT or its product, DDE, the fish in that water were polluted, too, and the pelicans ate the fish. Result: Eggs without a shell, or with so soft a shell that there was no chance of its being incubated and hatched. So there was no new generation coming along to replace the birds that died by accident and old age.

What was different in Florida? The answer seems to be that even before DDT was outlawed, there was less of it used where it would drain into the waterways that ran to the feeding grounds of the pelicans. The percentage of it in the bodies of the birds was often low enough to allow an eggshell hard enough to hatch and produce young.

Nevertheless the number of brown pelicans in Florida has decreased. Researchers claim that the number hatched in twenty or more wild rookeries—great stands of mangrove trees filled with nests, that cover islands and shores—is more or less stabilized. But something is happening to the birds after they are old enough to fly.

RAISING A PELICAN FAMILY

After a pair of pelicans have mated, they build a nest, placing it near the top of a mangrove tree in the rookery or somewhere along the shoreline. Other nests may be quite near it, but farther away than the pelican can reach when sitting on its own nest. The male brings twigs and sticks and presents them to the female, and she makes the nest, rounding it out in a big bowl. Sometimes she adds moss or other soft materials to the center. Later on, both birds bring additional soft materials, including their own downy feathers, for the nest's center. Both raid any nests in the neighborhood that are left unguarded, carrying away as many sticks as they can pull loose.

Normally two or three eggs are laid. The young hatch in about five weeks after the eggs are laid, the eggs having been kept warm by both parents taking turns. Their six-foot wing-spread gives them plenty of cover for the eggs. The heavy bodies, lowered to the eggs, would seem to be enough to swamp

Young pelicans on their nest

them. But the eggs are rarely injured unless they have softer shells than is normal.

The young are first pink-skinned, then gray, and naked of feathers when first hatched, when they are especially susceptible to hot rays from the sun. But they shortly acquire a covering of down that helps to keep them cool in the sun and warm on chilly nights. To feed them, the parent birds fly to the edge of the nest and regurgitate fish they have swallowed, smearing it on the bottom of the nest from which the young ones pick it up. As the babies grow larger, they learn to thrust their bills deep into the pouch of a parent bird, where they scoop up regurgitated food.

COLOR CHANGES

Gradually the young pelicans get the brown feathers that give them the overall coloring of the adult. Throughout their lives, brown will remain as the color of the great, six-foot wings, the back, and the breast. But the head and neck change radically in color from time to time.

In the breeding season the top of the bird's head is bright yellow, and the back of the head and neck are clear white that extends to the shoulders. A brown area runs from the eyes down the front of the neck. During the nesting, these markings change dramatically, to a white head, a white band running

This adult pelican is clearly marked with brown and white.

down each side of the neck, a white patch on the throat, and brown elsewhere on the neck. These markings are most noticeable in the older birds.

YOUNG PELICANS ON THEIR OWN

When the young ones are about three months old, they are ready to leave the nest and begin fishing for their own food. Very seldom are they fed by the parent birds after they leave the nest. They must learn the technique of poising in the air over the water, pulling in their wings, and diving headfirst to catch fish in their bills.

When a bird comes up with a fish, the fish is not immediately apparent; it will be in the pelican's pouch, along with two to three gallons of water. The pelican sits quietly, its bill tilted downward, and lets the water drain out. Then it flips its head up and the fish into its bill, and swallows it. Often a gull is nearby, to grab the fish if possible. If all goes well and the fish goes down the pelican's gullet, the pelican will wiggle its tail as if in triumphant announcement. But often the gull, sitting close, even on the pelican's head, is able to make off with the fish, and there is no happy wiggle.

This whole process must be mastered quickly, partly by instinct, partly by imitation of the older birds, to allow the young pelican to eat after it has left the nest. Some researchers believe that a large percentage of each year's crop of young starve to death before they learn to fish for themselves. If a pelican survives this first dangerous year, it may live from five to twenty years or longer.

WHY FLYING PELICANS ARE IN DANGER

Even so, danger awaits the flying pelican, and its death rate seems to be increasing even in Florida, where the numbers of

Even when it can fly, the adult pelican is not out of danger.

young hatched are more or less stabilized. As we have seen, a common sight there in the past was a flight of ten or more pelicans, over water or over land. At the fishing piers, every post and pier was topped with a sitting pelican, waiting for the boats to come in. Today there are not all that many. The piers are no longer mobbed by them; formations in the air are not as common.

Observers agree that the biggest threat today to the flying pelican is the fisherman and his hooks. Those who work with pelicans estimate that of all the birds brought in for help of one kind or another, at least 85 percent have fishhooks in them. Usually a leader and often a length of fishline dangles from the hook, making it easy for the pelican to become entangled in the mangrove trees where the birds nest and roost. Saddest sight of all to a pelican researcher is that of the remains of a bird dangling in the trees from a length of fishline.

We watched one morning at a sanctuary where a pelican was brought in with several hooks in him. In all there were five— two in his pouch, one in his throat (it narrowly missed the windpipe), one in a wing, and one in his back. The latter was discovered only when an attendant ran her fingers all over the bird, under the feathers, an investigation that is always performed and usually locates additional hooks.

The operator cut off the barbs of the hooks with heavy wirecutters, and so was able to slip the hooks out backwards. In a matter of minutes the pelican was free of hooks for probably the first time in his adult life. Then each puncture was sprayed with an antibiotic, and the bird was put in a holding cage. After a few days, if he displayed no evidence of infection, he would be released and allowed to go free.

The operator of the Suncoast Seabird Sanctuary told us that this was just an ordinary day—that on the average, probably more than one bird was brought in each day for this kind of treatment.

This accidental kind of danger to the big birds is bad enough. But the operator told us, with flashing eyes, of acts of deliberate cruelty where he had seen the consequences: Tying the bird's feet together and leaving it helpless in the mangroves, shooting firecrackers at it, shooting with guns, and a variety of other acts aimed at the torture and ultimate death of the pelican involved. A heavy penalty is lowered on anyone caught at this kind of mischief—but the perpetrator is seldom caught.

OUTLOOK GETTING BRIGHTER?

In spite of the many people engaged in protecting the pelicans and rescuing them from the troubles that seem to follow them, the status of this interesting species of bird is very vulnerable. For a time, the outlook for the pelicans seemed quite a lot brighter than it did before. Louisiana imported, in 1968, 50 young pelicans from Florida, nine to twelve weeks old. These were divided between two centers. The pelicans at one eventually died out, while those at the other, Rockefeller Refuge on Grand Terre Island, prospered. This was followed up with 415 additional pelicans transplanted in the next few years.

Nesting soon began and was repeated each year, with the number of young increasing each year. Eggshells seemed to have about the same thickness as those in Florida, and the young birds were fed until they could fish for themselves.

These conditions led Louisiana researchers to predict that "our official state bird may once again be a common sight along sections of our coast."

Then, in the summer of 1975, the pelicans suddenly began to die, until there were only about 100 left in the Louisiana colony. Their deaths were traced to the insecticide endrin, but researchers could not agree on where they got the endrin. Some thought it had come down the Mississippi River from cotton and other croplands, where it was used to replace the banned DDT. But the only agreement that investigators seemed able to arrive at was a great lack of information. In any case the pelicans died.

In Texas the picture has not seemed to change much. There were approximately 125 brown pelicans on the lower Texas coast in the fall of 1974, according to the Texas Parks and Wildlife Department. Most of these, however, were believed to come from Mexico. In 1974, there were fourteen known brown pelican nests in Texas, which produced one chick. The next

A rookery in the mangroves where pelicans, egrets, and other birds have nested

These pelicans have made their nests on the ground.

year saw eighteen nests, but records concerning them are incomplete.

California seems to be having a happier story. There is one nesting colony in California, at West Anacapa Island in the Channel Islands National Monument; it overflows onto nearby Santa Cruz Island. As recently as 1970, this colony made 552 nesting attempts but produced only one bird. Investigators found, in the nonproducing nests, addled eggs and broken and thin-shelled eggs not capable of producing and hatching young birds.

Decreasing use of DDT, with a ban on all its major uses, has seen an improving hatching record here. In 1972, 57 young were produced by the colony, in 262 nests. In 1974, 416 nests were counted on the two islands, producing 305 young birds. The colony's success continues with every passing year.

It would seem that the danger of the pelican becoming extinct is ebbing. However, a bad year where the bird is now gaining could throw the situation into catastrophe. It is possible, then, since the pelican is so vulnerable, that it will be on the endangered list for some time to come—a situation clearly brought about by the contamination of its home areas by pesticides.

Chapter 2

LIVING IN MANY
KINDS OF HOMES

"Is that coon out there again?" the national park naturalist asked with a chuckle. He was giving a lecture on the plants of the Everglades, but most of his audience was looking toward the back of the room instead of at him. Sure enough, a small raccoon stood in the doorway, looking for all the world, with the black mask across his face, like a small, furry bandit. For a full minute he boldly returned the gaze of the delighted audience, then ambled off, and the naturalist and audience, including us, returned to the lecture.

Raccoons are intelligent as well as inquisitive animals, and this little fellow had apparently learned at an early age that he was safe among the people in a national park, where no one is allowed to harm the animals. So he could satisfy his curiosity without fear and sometimes might even get a handout.

A POOL WITH WILDLIFE

A few days later, we were driving on a dirt road along a small stream in the Everglades, and came to a place where the stream widened out into a pool. It was a time of severe drought, and the pool was one of the few watering places left in the area. So it teemed with wildlife.

Alligators of all sizes swam in the water or lay asleep on the bank. Turtles sunned themselves on logs. Blackbirds whistled among the marsh plants that grew along the edge of the water. Birds of many kinds perched in the trees around the pool or waded on long legs in the water, catching frogs or fish. At the back of the pool a trail ran down through the grass and brush to the water. It had been made by deer and other animals that came to the pool to drink.

We stood quietly at the side of the pool and watched the animal life. Suddenly we saw a mother raccoon and three young ones walking one behind the other, making their way through the brush above the pool. A fourth baby raccoon, with the raccoon's natural curiosity, stopped on the trail to watch the other animals and us.

An otter slid down the bank into the water and came up with a large fish, still alive and flopping, in its mouth. It laid the fish on the bank and disappeared.

All at once a loud noise started. The otter had slipped through the brush and grabbed the baby raccoon, dragging it down the trail toward the water. If it could drown it, the young raccoon would make a good dinner. The raccoon was screaming loudly, and the birds also set up a great clamor.

We watched in horror, too surprised for a minute to do anything. The two struggling animals entered the water. Then one of us picked up a rock and hurled it at the otter. The rock landed with a splash close to the otter. It was so startled that it let go of the raccoon and the raccoon quickly swam out of the water and dashed back up the trail. There it joined its mother, who had come back to help her baby.

The otter bared its teeth at us, then swam over to the bank where it had left its fish. But in the meantime the fish had flopped back into the water, and the otter left, without any dinner that day.

A RACCOON CAN ADJUST

Unlike many animals, the raccoon has been able to adjust to a changing environment. For that reason, it flourishes in almost every state in the Union, and we have found it in every area we have visited—along roads and trails in the Great Smoky Mountains; in campgrounds in Florida forests; in the North Woods; in the Ozarks. One "washed" our dishes on a campground table in Oklahoma by licking them off, one by one, and throwing them on the ground; one took the whole lump of suet we had put out for birds at our Indiana home; one led her new family down the road ahead of our car in a park in Illinois.

It has been believed that there were no raccoons in the Rocky Mountains, but they were prevalent there when we visited Colorado in the summer and fall of 1976.

The raccoon is distinguished by a black mask across its face and alternating black and white rings on its bushy tail. Its thick, long hair, a mixture of gray and brown shading into

Raccoons frequently find refuge in trees when they are threatened.

Two baby raccoons

black, is used for fur coats, fur collars and trim, and other articles of wearing apparel. Pioneers, such as Daniel Boone, wore coonskin caps, with the tail hanging down at the side.

Raccoons vary in size from twenty-five to thirty-two inches long, with the tail an additional nine to ten inches. In weight, they are twelve to thirty-six pounds. A very large one might weigh forty pounds or more, especially in the Northwest where they are the largest. They are smallest in the South, and in the Florida Keys some may weigh as little as three to six pounds.

Baby raccoons weigh only 2½ ounces at birth. They are covered with a soft, fuzzy coat and are born with their mother's black mask. They do not open their eyes until they are about three weeks old. There are usually about four in a litter, but may be as many as six or seven or as few as two. When the babies are about two months old, the mother begins taking them on foraging expeditions, short at first, but longer and longer as the young ones gain strength and experience. Raccoons are mainly nocturnal, coming out at dusk and heading for their home nest at dawn. Sometimes, however, they venture forth during the daylight hours, as we saw them at the Everglades pool.

For habitat, they prefer a wooded area with a stream nearby, where they can find crayfish, their favorite food. However, they can make a home in almost any kind of habitat, and that is why they have survived in our changing environment in much larger numbers than other animals that are less adaptable and are on the endangered list.

A raccoon's favorite place to make a nest and have its young is in a hollow tree, but it often uses whatever comes in handy— a deserted burrow in the ground or one it has taken away from a weaker animal, a deserted hawk's nest, a deep hole in a rocky ledge in a mountain or desert area—anything, in fact, that offers a haven.

A RACCOON "WASHES" ITS FOOD

The Latin name for raccoon is lotor, meaning "the washer," so-called because of the coon's peculiar habit of "washing" its food if it is near water. It holds the food underwater with its front paws, which are much like a person's hands, and washes it with long, slender "fingers." Scientists have not agreed on why the coon does this. Some think it learned to do it by washing the sand off of crayfish and other seafood. Others say that maybe it just likes the feel of water on its paws. Another theory is that it dunks its food in water for somewhat the same reason that a person dunks a doughnut in his coffee. Our personal feeling, though unsupported by scientific research, is that the little rascal does it just for fun!

A raccoon eats practically everything, including all sorts of berries and nuts, bird's eggs and the baby birds, small rodents, lizards, toads, and frogs. Garbage is high on the food list of this little animal, and many a camper has heard the clang of the campground's garbage pail lid, as the midnight marauder helps himself to the remains of the camper's supper.

A delicacy for raccoons near the ocean is turtle eggs. When a green turtle or some other giant sea turtle comes from the water to a sandy beach and lays her eggs in a hole she digs for them, a watchful coon is likely to dig them up and eat them as soon as she has turned her back. That is one of the reasons these turtles are on the threatened list.

The raccoon sometimes raids the farmer's poultry yard and eats the eggs and chicks and, occasionally, the adult chickens. It is especially fond of sweet corn and will sometimes damage much of the crop in the field, tearing off ear after ear of corn and leaving them half eaten on the ground. So it has earned the enmity of the farmer, who carries on unremitting warfare against it in some parts of the country. The raccoon is often killed for its flesh, too, which many people consider fine eating.

A raccoon digging snails and other food at the edge of a marsh

A RACCOON'S DEFENSES

When threatened with imminent danger on the ground, the coon often takes to a tree, which it climbs with the greatest of ease, coming down frontwise like a squirrel, or backwards. When it is hunted with dogs and men with guns, going up a tree is an unfortunate move, however, for the men can quickly shoot it down. When pursued by only one or two dogs, the coon may turn ferociously on a dog and injure it severely or even kill it. The story has been told of the raccoon that led its pursuer into a stream and then climbed on top of its head and held it underwater until the dog was drowned.

In its nightly wanderings for food, the raccoon may travel for several miles, but dawn usually finds it curled up in its nest,

sound asleep. During cold spells, it may sleep for several days, but it does not hibernate like the bear, which sleeps all winter.

The raccoon is a loner. Except for a mother with babies, it ordinarily travels alone and lives alone. One condition that may draw a group of them together is an abundance of food concentrated in one spot. A man in Florida put a long trough behind his restaurant and filled it with scraps from the tables, and coons came from all directions to eat the "goodies." It was a sight to see them crowding around the trough, and they got along together all right and there were no fights, at least as long as we watched. Our brother in Indiana bought dog food for the coons in his neighborhood and put it out on the ground at night under a big window, which offered a grandstand seat for seeing them come out of the surrounding woods—thirty or more. A little opossum also came timidly to the party. Again we saw no fights, although a mother raccoon ordered her young ones up a tree when she apparently thought there might be trouble in the crush of animals around the food.

So, with the raccoon able to make itself at home in a great variety of surroundings, we find it almost everywhere we go in the United States and southern Canada. It has almost no rival among the animals in its ability to make use of the food and home sites wherever it finds itself.

THE LAST ANIMAL

Somewhat like the raccoon—even surpassing it—in the way it can live in many kinds of homes, is the coyote. Indian legends name the coyote as the animal that "will be the last animal on the earth." It lives almost everywhere—in nearly every state in the United States and in most of Canada and Mexico.

Man's hand has never been turned against any other animal as it is against the coyote. Through the years great drives have been staged to wipe it out in the interest of protecting livestock, and thousands have been killed in a single drive. Thousands more have been poisoned and trapped; as many as 125,000 a

A mother coyote brings a dead rabbit to her pup at the den.

year have been killed. Yet the coyote survives and continues to prosper.

What kind of animal is this that can stand up to such abuse? The coyote looks something like a medium-sized collie dog; it is about four feet long, with a pointed muzzle and pointed ears and an alert, inquisitive expression. Its tail is round and thick, and is often carried low as it travels. It is grayish, with buff-colored legs, white beneath, and black-tipped ears and tail.

Probably most known for its personality, it may be our most resourceful animal, cleverest in avoiding enemies and in getting food. Its enemies include the larger carnivores—bears and mountain lions; the members of the deer tribe such as moose, elk, and deer, can sometimes trample a coyote to death if it attacks them. But its major enemy is man. The coyote has been known to dig up traps, no matter how carefully they are buried, and to avoid carrion in areas where it contains poison, and in other ways to defeat the men who are always after it.

Coyotes often work in pairs or in larger groups stalking and catching food—rabbits and squirrels and other rodents. And often a single coyote relies on its speed, as much as forty miles per hour, to overtake a jackrabbit or other prey. Sometimes a coyote will find a badger digging out the den of prairie dog or ground squirrel, and will wait in the vicinity to grab the prey the minute it shows above ground at one of several entrances. The coyote seems to have no conscience about stealing the meal of the badger, with which it does not share the catch.

THE COYOTE'S FAMILY

Coyotes often mate for life, and dig their den together. Almost any shelter may be taken over by them—dens of other animals, hollow logs, caves in rocky formations—whatever will protect them from weather and enemies. They can dig in somewhat loose earth very effectively, and if making their den in such a place, will run a tunnel up to thirty feet long, a foot to six feet underground, and enlarge a little chamber at the end, which is the den.

Here the pups are born, in a litter of five or six on the average but sometimes much larger; one of nineteen is on record. But sometimes there are only two or three. The pups, born in April to June, are covered with brown, woolly fur. They are born with their eyes closed, but the eyes open in ten days to two weeks. They are cuddly-looking little things, with sharp, upstanding ears, and seem to be full of fun.

By the time they are about six weeks old, they spend the daylight hours around the mouth of the den, playing with each other and teasing their mother. She may, however, be hunting for food for them; and the father is almost certain to be off hunting. While the pups are being born and for the first weeks of their lives, he lives in a separate den but is constantly in attendance to bring them food. To wean them from her milk and introduce them to solid food, the mother regurgitates food that she has partly digested. When they are about two months old, they are taken abroad by the parents and the process of learning to hunt is started.

WHAT A COYOTE EATS

Besides their meat diet—mostly rodents—coyotes eat many fruits and some vegetables. They are especially fond of watermelon, and can easily find the ripe ones, taking a bite or two from each and so damaging a large number in a field.

But it is damage to livestock for which man most blames the coyote. If lambs are killed, the immediate assumption is that the coyote is responsible, and sometimes it is. But often dogs and other animals are to blame, and the coyote gets far more than its share of the efforts to kill the culprit. However, in recent years this situation has shown a tendency to reverse itself, and man seems to be relenting in matters of state-paid bounties and the wholesale poisoning of coyotes. There is a national ban against poisoning coyotes on any land that is government-owned.

It has been shown that the coyote keeps down the population of crop-destroying rodents and so often more than pays its way. And it has been found that control can be achieved through eliminating those animals that have become habitual takers of lambs, rather than trying to destroy the entire coyote population.

A Colorado biologist has gone on record with the view that there should be a base population of 50,000 coyotes in that state. The animal will have no trouble maintaining its numbers, aided by the lessening pressure of man and by its own almost unbelievable adaptability. The coyote seems able to thrive in almost any surroundings, city or country. Several thousand are believed to live within the city limits of Los Angeles, holing up in culverts, rock outcroppings in hilly sections, and in burrows dug into the earth.

THE "SONG-DOG"

Presence of the coyote is readily noted because of its very distinctive bark, especially on moonlit nights—a sharp yapping followed by a long-drawn wailing howl. There are many variations, and together they have earned various nicknames for their maker, such as "song-dog." A single animal is usually making all the noise, although it sounds like several together, and the sound seems to come in from all directions.

People who live near the coyote agree that it seems to have a sense of humor. A cowboy that we knew on our Colorado ranch told of riding at night, whistling to his horse in time with its walking, and gradually whistling a little faster to get the horse to walk faster. He made good progress, building up speed, until suddenly a coyote would howl nearby. At the sound, the horse would immediately break into a trot or gallop. If the rider wanted to go back to a fast walk, he would have to start all over again. The coyote seemed to be going along with him, because again and again it broke up the smooth, rapid walk with its ear-splitting "song."

Another example of the coyote's sense of humor has been seen recently on a golf course at Estes Park, Colorado. Many golfers playing there have made a long drive on one of the fairways, only to see a young coyote follow their ball, grab it, and make off with it.

Authors of books who write about coyotes often wind up their account by expressing the hope that this animal will not disappear, especially in the West. Its disappearance does not seem to be very likely—indeed, it would seem that the Indian legend is close to the truth, that the coyote will be the last animal on the earth.

Chapter 3

PROTECTION MAKES THE DIFFERENCE

We sat in the boat, ten of us, a few hundred yards from a big mangrove island in Everglades National Park. Thrilled and fascinated, we watched hundreds of birds in the tops of the mangrove trees on the island, where they had built their nests high in the trees. The boat brought park visitors here so that they could see the birds—the famous big birds of the Everglades.

For an hour we had ridden over sparkling water in the bright Florida sun—up rivers and through canals and lakes, and finally into this big lake. We would go no nearer the rookery, so as not to frighten the birds. But we had a fine view from here, because each of us was looking through a pair of binoculars. We could see the birds as well as if they were very near us.

There were hundreds of white egrets—American egrets and the smaller snowy egrets. There were big brown pelicans, and brownish-looking cormorants, and gold-to-black anhingas. The island was alive with dazzling, flickering white and black and golden wings, as the birds flew in and lit, fed their young, and flew off again. Many of them hovered over the trees for a moment or two, as they came and went.

Each pair of birds had its own nest. There were eggs in some of the nests and young birds in others. On most nests a parent

Limpkin watching for his dinner in the murky waters beneath his perch

bird was sitting—perhaps the father bird, perhaps the mother, taking turns at brooding the eggs or sheltering the young. No pair of birds left the next unattended for very long. If the eggs or the young birds were left alone, gulls or crows might swoop down and grab them, or a snake might eat them. And before the young birds' naked skin is covered with feathers, the hot sun can burn them enough so that they die. We saw some of the parent birds standing with wings spread over the nest, in order to shade the babies from the sun.

What a racket the young birds were making! Whenever a parent came in with a fish or frog for its young, all the young neighbors set up a loud screaming and squealing. So, as food was arriving all the time at one nest or another, nearly all the babies were screaming most of the time.

Many of the big birds in the Everglades nest in island rookeries like this one. Some kinds mix with colonies of other kinds. Some keep their colonies apart.

THE GREAT WHITE HERON

Largest of them all, the great white heron is a bird that keeps its rookery apart from others. It doesn't seem to care for too much company at any time, and only two or three pairs or sometimes up to a dozen or more will make their nests together. Their nests are platforms that they build from twigs and branches in the tops of the mangroves.

This big bird stands more than four feet tall, and when it opens up those long, powerful wings, they spread to more than six feet from tip to tip. It has greenish-yellow feet and legs and a yellow bill.

The great white heron is most often seen standing or wading with stately tread in the shallow waters of Florida Bay, where the Great White Heron National Wildlife Refuge is located. It feeds on fish and crustaceans and other seafood, which it spears from the shallow water. It is protected by law throughout Florida; this and neighboring refuges give added protection to its rookeries and feeding grounds, giving it as much home area as possible in its natural condition.

This striking bird is very much in need of protection. It is easily frightened away from its nest, even leaving young birds, which themselves were once hunted for food. It is often the victim of hurricanes, which, with illegal hunting, once dropped its number to little more than 100. Today the count seems to be around 2,000, of which 1,500 are in Everglades National Park.

Another likelihood of loss seems to be that it interbreeds with

Great white heron

the great blue heron, and so the pure strain of the great white may disappear. Special effort is being made to maintain the Florida Bay areas to which the range of the great blue heron does not extend.

THE HERON FAMILY

There are more than sixty members of the heron family, scattered all over the world. As well as the herons, all of the egrets belong to this big family.

Members of this family are alike in many ways. Most of them have a long, oval body and a limber neck that can be extended straight up or out from the body, or curved in an S to bring the head low. When they fly, the neck is held in the S position, and the feet trail out straight behind the body. They have long legs

Great blue heron

and strong feet, and a bill that is straight and sharp-pointed.

Heron-family birds usually live near the water, in marshes and around the edges of ponds, lakes, and bays. They feed on water life along the muddy edge, eating fish, crayfish, crabs, grasshoppers, water insects—anything that moves there.

Many of them nest in rookeries, and they often roost together at night, on mangrove islands or in trees on the mainland, in "communal roosts." Their cry is harsh and guttural, and very loud in some of the larger kinds. Many of them have courting plumage in the mating season. They build a rough platform of sticks as the foundation of their nest and line an inner nest with moss and leaves.

We have more than a dozen of the heron-family birds in the United States, and all these kinds live at one time or another in the Everglades. They include, besides the great white and the great blue, the little blue heron, the little green heron, the cattle egret, the Louisiana egret, and others. Some of them besides the great white heron have been threatened with extinction, and their species has been saved through man's protection.

TWO COUSINS

The elegant little white heron we call "snowy egret" stands only about two feet high. It is so beautiful, with its dazzling white feathers fluffed and glossy in the sun, as it moves slowly, gracefully, in search of food, that it steals the show from its much larger cousins. It has bright yellow feet, made even more striking as they appear at the ends of long, black legs, and so its other name—"golden slipper bird."

Look for the snowy along the edge of shallow water, at ponds, lakes, and marshes, either salt or fresh water. It will be wading along, shuffling its feet and so stirring up the muddy bottom, and darting after a fish, crayfish, crab, shrimp, grasshopper, or water insect. Often it spreads its wings a little as it moves. Watch it using those strong feet as it perches in a bush or tree, and notice how its toes hold onto a branch as it actually climbs up or down on it.

In late winter and early spring, before they mate and make their nests, the snowies become even more beautiful. Both male and female birds grow handsome "courting plumage"—long, lacy plumes that the birds can cock high on the back of head and neck, and that trail from shoulders and back. As many as fifty of these graceful plumes grow on a single bird.

Before they mate, a pair of egrets perform a courtship "dance." They bow low to each other, and parade around each other in slow circles, the lacy white plumes cocked high and trailing almost to the ground from the sides and tail.

When the courtship is over, the pair fly to the rookery and build a nest. It is a flimsy platform of dry sticks, perhaps built on the remains of a nest from last year. It may be in a tree, or it may be in marsh weeds close to the water. The nest itself is lined with moss and leaves.

Here the mother egret lays four or five oval, pale blue-green eggs, about 1½ inches long. The eggs hatch in about three weeks, to release scrawny, naked, completely helpless babies that move only to lift their heads and open their bills wide and scream for food. Now the parent birds must work throughout

Snowy egret

American egret

the day, taking turns, to bring food to these ravenous young-sters. This goes on for about six weeks, when the young birds are large enough to fly, to leave the nest, and to find food for themselves.

All through the summer and fall the egrets, parents and young alike, have a fine time around the sloughs and lakes and ponds and on the islands. Some of them leave the Everglades and fly north, a few going as far as Canada. In the fall they return to the South. In winter and early spring the snowies again pair off, as their courting plumage grows, and by mid-spring they are raising new families.

The snowy is one of two cousins in the heron family that are very much alike. The other is the common, or American, egret. It is a little larger than the snowy, and instead of golden slippers, has black feet and legs. Its bill is yellow.

It lives very much like the snowy, and eats fish and crustaceans and other seafood that it stirs up from the muddy bottoms of sloughs and ponds. And like the snowy, it grows courting plumage in the spring, before starting to raise a family.

THE PLUME BIRDS

These two are important members of the "plume birds," a famous group of birds that won man's protection and survive because of it.

Early in the century, the beautiful plumes of the egrets and of some other birds were used in great numbers to decorate women's hats. The plumes brought fancy prices from milliners, and the birds were slaughtered by the thousands to get them. Not only were the parent birds killed, but many nests were left with eggs and young in them.

It seemed as if nothing could save the plume birds from extinction. Laws were passed against killing them, but poachers continued to slaughter them. Finally, a warden was killed when he tried to arrest some poachers. Then the country rose up against the use of the plumes in millinery, and laws were passed against it. So there was no market for the plumes, and killing the birds was effectively stopped. It was almost the first thing like this that had happened in our country; protection of the plume birds helped to mark the beginning of conservation in order to save a species.

Roseate spoonbill

ROSEATE SPOONBILLS

Another important plume bird is the roseate spoonbill. It is not a heron, but belongs to the family that contains the white ibis and the wood ibis.

We first saw many spoonbills together when we were driving from Miami, Florida, to Everglades National Park. We stopped alongside the road, where many other cars had also stopped, to watch a flock of spoonbills feeding in a slough near the road. Something startled them, and they flew, rising steeply and wheeling over us. As we watched, looking up, we seemed almost surrounded by a rosy cloud, and that impression remained, coming from the rosy undersides of so many of them together. It was quite a while, as we continued our trip to the park, before the feeling of a rosy cloud left us.

We saw spoonbills again from a boat on a lake in the park, where we watched their rookery. This was a rookery solely of spoonbills; they do not seem to like joining with other birds to make their nests. They went and came, two or three or more at a time, to feed their young. Sometimes they flew directly over us, and here again was the rosy cloud. Their most brilliant coloring is in the rosy feathers underneath, although they also have shoulder markings of rose and red. They are white above except for these markings, and the young birds are more white than rose underneath.

A more common way of seeing spoonbills is in the ponds along the park roads, where any number of one to half a dozen ordinarily may be seen feeding. They wade along through shallow water near the edge, the long bill, with its end like a spoon, thrust down into the water. As they move, they push the bill back and forth through the mud of the bottom, stirring up small crustaceans that they catch in the "spoon."

They are seen commonly in southern Florida, especially in Everglades National Park, and sometimes farther north in Florida, and sometimes along the Louisiana coast. A large colony of them on the Texas coast near Smith Point is protected by the Audubon wildlife sanctuary, Vingt-et-un.

PROTECTION HAS SAVED THE SPECIES

With all of these birds, protection by man has been the important thing in saving the species from extinction. In the great white heron, it is protection of its home surroundings. In the plume birds, it is protection from being slaughtered for market by man himself.

Assured of a home undisturbed by changes and by violence, these birds have come back from the very edge of extinction, and so live on to be enjoyed by all who see them.

Chapter 4

THE SQUIRREL COUSINS

Snow had come down and come down, until it stood six inches deep on the floor of the meadow. It had drifted heavily around the trees and big rocks and against the banks of the little creek.

The bobwhite quail—eight of them—had found a place where they were protected from the storm. The lower branches of a pine tree were close to the ground, and snow falling on them had weighed them down and made their tips touch the ground. Snow covered them and made a roof. Under the roof was a little space—a shelter. It was like a small tent, and it protected the quail from the wind and falling snow.

The quail came to it along a little path they had made, walking single file through the snow and beating it down in a narrow trail—a "quail trail." They had gone in and out along their trail several times, going out in search for food and coming back to their snug little shelter in the evening.

Now light of a new day was just coming and bringing danger to the bobwhites. Their warm little shelter was turning into a trap.

The quail were still asleep inside it. Outside a hungry bobcat was moving quietly along their trail. He had hunted through the night but had found little, and his empty stomach was demanding food.

He sniffed eagerly at the strong smell of quail, and came

Bobwhite quail in a clump of grass; male at left, female at right

closer and closer to the sleeping bobwhites. Quail for breakfast! That was just what he needed. Carefully he moved along, not making a sound. If he could find them before they saw him, he could make an easy kill. He could see where the trail went under the branches. He was almost within reach of them and still moving slowly, soundlessly.

Everything around him was just as quiet. It was as if the whole early morning was waiting for the bobcat to spring.

Then a racket began that seemed to split the air above him. It seemed to be everywhere, but it came from over his head.

Tcher-tcher-tcher-tcher-tcher-r-r-r! Tcher! Tcher! Tcher-tcher!

It went on and on. The bobcat stopped still in the path and looked up into the tree beside him.

On the trunk of the tree, about ten feet over his head, a red squirrel was stretched out, head down. The sharp claws on all

Bobcat

four of his feet clung firmly to the tree trunk. His tail whisked about wildly, this way and that way, over his head, then lay flat against the tree. His head was lifted, and he was peering down at the bobcat.

And he was scolding with all his might. He had caught a bobcat creeping along after something, and he was yelling as long and loud as he could yell.

He was so angry he could not sit still. Even when he was spread out on the tree trunk, his whole body jerked with his scolding. In a moment he turned, whisked up the tree a little way, and sat up on a branch. His scolding never stopped for a second, and his tail never stopped jerking.

In their shelter, the bobwhites were stirring, waking up. The cat heard them moving and saw one or two of them. Without waiting any longer, he quickly ran a little closer, and pounced. If he was lucky, he might get one of them.

But they had been well warned. As the cat moved, they flew.

Some of them went out the entrance, some went through the snow along the sides. There was a small explosion of quail, sending the snow flying. They flew into nearby trees, and the cat missed them all. All he got was a face full of snow, which he didn't like.

He sniffed wistfully around the little shelter, the quail smell making him hungrier than ever. Then he went back to the tree where the red squirrel still scolded. He stood on his hind legs and stretched upward along the tree trunk. The squirrel stayed over his head and scolded harder.

The cat could climb the tree, and perhaps he considered doing it. Squirrel for breakfast would be as good as quail. But if he started climbing, the squirrel would dash higher, three or four times faster than the cat could climb. If the cat followed him to the top of the tree, the squirrel would run out on a limb and jump to the limb of a nearby tree.

The bobcat knew better than to waste time with this kind of game. So he dropped back to the ground and went back along the path, the way he had come.

Perhaps he could find a mouse or a rabbit for his breakfast. But first he must get far away from that red squirrel. If he stayed here, he would go hungry, because the squirrel was still warning everything within hearing distance.

CHICKAREE — BIG CHATTER

The red squirrel is a born scolder. He is a small squirrel, but he scolds longer and louder than any other animal. If a person or a bear or a dog walks through the woods, the red squirrel scolds. He gets so excited that he drums his hind feet on the trunk of the tree. He jerks his tail. He dashes to a higher spot, scolding all the time.

He chatters so loud and makes such a racket that he has been

Red squirrel eating a nut

given a nickname. His nickname is chickaree. It means "big chatter."

AT HOME IN THE TREETOPS

At the slightest alarm, chickaree makes for the nearest tree as fast as he can go. Once in it, he is right at home. He runs up a tree trunk at top speed, and he can come down just as fast, head first. Going up or down, he can run around and around the tree trunk.

He flashes in and out among the branches. Swiftly, he runs along a branch, to the very ends of slender twigs. From them he leaps to the tips of the branches of a nearby tree. On he goes without the slightest pause, running like mad.

Chickaree lives in the treetops more than he does on the ground. When he travels from one tree to another, leaping from branch to branch, he has a regular route that he follows. He knows this route so well that he can travel along it at top speed. His back, sides, head, and tail are a beautiful russet red, and he looks like a bright streak as he runs, especially in sunshine.

He can run and leap because of the way his feet and legs are made. His claws hold to the bark of a tree trunk and to the branches of trees. His hind legs are so strong that he can jump a long way. His tail helps, too. He uses it to balance himself as he runs and leaps.

A SQUIRREL'S ENEMIES

Chickaree is small, and he is good to eat. So every meat-eating animal in the woods is his enemy. Bear and fox, weasel and bobcat, hawk and owl—all of these and others, too, find the red squirrel fine eating—if they can catch him. But he can get away from most of his enemies by running up a tree.

A weasel or a bobcat chasing him can climb a tree. Chickaree can outrun these animals, and he can run through the treetops where they cannot go. He can twist and turn and dodge and jump until he gets away.

Another enemy of the red squirrel lives in the big evergreen forests in the north. He is the marten. A marten can turn and dodge and jump, too. He can follow chickaree almost anywhere, and he can move as fast, or faster. So martens often catch and kill red squirrels.

A fox can creep up on chickaree on the ground. But if the squirrel sees the fox, he is up a tree like a flash. Then he sits on a limb over the fox's head and scolds with all his might.

If a hawk comes near, chickaree flattens himself out on a tree trunk. Maybe then the hawk will not see him. Or maybe the hawk will swoop at him there.

Then the squirrel's tail helps him again. When he is flat on the tree trunk, his tail is the biggest part of him. The hawk grabs at the fluffy tail, but he cannot hold onto it. Like a flash, the squirrel darts around the tree and is gone.

Chickaree often comes to the ground to look for food, and he often stays on the ground to eat what he finds there. He is likely to be in danger there, and he knows it. He keeps a close watch for any enemy that might be stalking him.

He has good eyes for watching; he almost has "eyes in the back of his head." They are on the side of his head and near the top of it, bulging out a little at the widest part of his face—so they can see behind him, and above him, and all around him. He can see an enemy coming from any direction.

When he eats, he never keeps his head down for very long. He noses around on the ground for a moment until he finds something to eat. Then up he pops, to sit on his haunches and nibble the acorn or piece of corn he has found, holding it up to his mouth with his front paws. He is safer in this sitting-up position, because it gives him a better chance to see what is going on around him.

If he wants to see still more, he can sit straight and high, his front paws tucked under his chin. This is his "picket pin" position, when he looks like a peg driven into the ground. It gives him a fine view of all the ground around him.

WHAT A SQUIRREL EATS

Red squirrels can find something to eat almost anywhere in the woods or forest. They eat mushrooms and lichens, tender, growing twigs and buds, flowers and roots. More than anything else, they eat seeds. Much of their food is seeds from pine cones and the cones that grow on other evergreen trees. They eat the big, fat seeds that are nuts, like walnuts, hickory, and pecan nuts. They eat the acorns of the oak trees. They eat the smaller seeds that grow on many kinds of trees like maple and dogwood, and on bushes and vines like wild roses and bittersweet. Almost any kind of fruit makes food for them, too—blueberries and blackberries, wild cherries, elderberries, and many others.

When chickaree finds a nut with a hard shell, like a walnut, he cuts the shell away with his sharp front teeth. He cuts into a pine cone, shelling off the hard, outer scales so that he can get to the pine seeds under them. Then he cuts off the seed's shell, and finally he can eat the sweet, juicy kernel that lies inside.

He uses those four front teeth so much that they wear away at the sharp edge. But they do not get shorter, because they keep growing, and the new growth makes up for what is worn away.

By the time fall comes, the acorns and nuts and pine cones are big and fat. Chickaree goes up to the treetops and cuts them loose. Down they come, and he runs to the ground, picks them up, and carries them off. Some of them he eats then and there. Some of them he stores away.

Chickaree works hard all through the fall, storing food away. He works all day, and sometimes he works at night, too, carrying food to his storehouse. Then he must be very careful that a big owl does not see him and carry *him* away!

The red squirrel's storehouse may be a hollow log, or a hole in a tree, or a shed near a house. He makes a great pile of pine cones and acorns, called a midden. In winter he comes to his midden to get food. He throws shells and husks onto it as he eats, and so it gradually becomes a pile of trash.

Often chickaree buries acorns or nuts in the ground, one here, one there. He seems to remember where he has buried some of them, because he finds them again. In winter he digs them up and eats them.

But he never finds them all. Some of them begin to grow, under the ground. They are the seeds of the oak trees and the nut trees, and new trees grow from them. So chickaree is a planter of trees.

KEEPING WARM

Chickaree makes a nest in a hollow tree trunk. If he is lucky, he finds an old woodpecker's hole for his nest. He puts soft grass and leaves inside, and soft moss, if he can find it. Sometimes he chews up bark and puts it inside the nest. Anything that is soft and durable is likely to be something he will use. Sometimes he finds an old rag around a shed or a back door, and he will go to any amount of trouble to carry it away to put in the nest. When he has finished lining the nest with these soft materials, it is a snug, warm home.

Often he makes another nest on the branch of a tree. He carries twigs and leaves and trash up to the branch and makes a big wad of them, fastening the wad firmly to the branch. Then he hollows out the center and lines it with soft materials. This kind of home is snug and warm, too, but it probably does not feel as solid as the home inside the tree trunk. The squirrel seems to like a hole in a tree better, if he can find one.

Fall goes by, and winter comes, and chickaree curls up in his nest on many days that are cold and stormy. He wraps his tail around him to keep warm.

Yet on many winter days, he is out running through the tree-tops. He goes to his storehouse to get something to eat. He digs up acorns he has buried. If snow covers the branches, he can run on the underside of them. He can tunnel underneath the snow on the ground. When he sits up, he spreads his tail close along his back, and it protects him from the cold wind and the snow.

WINTER BRINGS GREATER DANGER

In winter he must be very careful. There are no leaves on the trees to hide him now. So he watches more carefully than ever for hawks and other enemies.

When there is an ice storm in winter, chickaree has a hard time. Every twig and branch is covered with slippery ice. So now the well-known pathways are filled with danger, and he moves much more slowly along them. At every step, he may slip. When he leaps, he may not be able to catch hold again.

Then he falls. Down, down he comes, all the way from the top of the tree.

Here again, his tail helps him. He uses it to flip himself over, so that he will land on his feet. Then he spreads his tail out as he falls, and glides a little, so that it breaks the force of the fall. Usually, when he hits the ground, he can run off as if nothing had happened!

SPRING IS A HAPPY TIME

At last the long winter is over. The snow is gone, the sun is warm, and spring is here.

Chickaree is livelier than ever in the spring. He goes racing through the treetops just for the fun of it. He chases other squirrels away, scolding loudly as he runs.

New green leaves are coming out now on the trees, and some of the trees have flowers. Chickaree nibbles at the leaf buds and flower buds. He goes to the very tips of the branches to get them.

There is another kind of food he likes in the springtime. Birds lay eggs in their nests in spring, and he sometimes eats the eggs. If the birds catch him near their nest, they fly at him. They would peck him to pieces if he did not run for his life!

BABY SQUIRRELS

Baby squirrels are born early in the spring, usually in the hollow-tree nest. At first they are very tiny and helpless. They have no fur, and their eyes are closed. Their mother feeds them with her milk. She keeps them warm through the chilly spring days.

They grow fast. Fur grows on their bodies, and their eyes open. Soon they come out of the nest and play in the warm sunshine. What a good time they have then! They tease their mother and chase each other up and down the tree and try to snatch food from each other. They are big enough now to eat seeds and mushrooms and fruit.

Summer comes, and they are almost as large as their mother. Soon they will be large enough to leave the nest for good and look after themselves. Soon they, too, will be racing through the treetops, and scolding their enemies, and storing away acorns and pine cones for the winter.

TWO WESTERN COUSINS

The red squirrel has two very close relatives that live in the West, in the big mountain forests. One of these is the Douglas squirrel, that lives in the states of the Pacific coast. The other is the pine squirrel of the Rocky Mountains.

Both of them are so much like the red squirrel that they, too, are called "chickaree." They chatter and scold and race through the treetops. They make two kinds of nests like those of the red squirrel. They store away nuts and pine cones and eat them in winter. And they look very much alike.

They are both more olive-gray than red. The Douglas squirrel is a rusty color underneath, where the pine and red squirrels are white. They are all about the same size, and all have a strong, black line along each side, especially in summer. This line seems to mark off a division between the dark back and the lighter underparts.

The Douglas squirrel adds a different noise from the usual racket of the red squirrel. Giant sugar pines grow where it lives. The green cone of a sugar pine is very heavy before the sap dries out of it. When the squirrel goes high in a sugar pine and cuts off one of these heavy cones, the cone comes down with a crash that can be heard all over the forest.

THE SQUIRREL FAMILY

The squirrels make up a large and closely related family of animals. The family's Latin name is Sciuridae, from a word meaning "squirrel" and also "fluffy tail." It is a good name for these animals, because all of them have hairy tails, and some have tails that are handsome, fluffy banners.

Nearly all the animals of the squirrel family come out in the daytime and sleep during the night, but they like nothing better than to stretch out in the summer sun and take a nap. They make nests in trees or under old logs and rocks, or burrow into the ground. Most of them run and leap, sit up on the ground to eat, and sit up like a picket pin to see what is going on around them. Most of them store away food for winter weather. They may have three to six babies in a litter, sometimes more; and some have more than one litter in a year.

There are many "cousins" in the squirrel family. Some of them are almost sure to be living near your home, in park or forest, in meadow or plains or hills or mountains.

THE TREE SQUIRRELS

Squirrels that run into the trees for protection, that are as much at home in the tops of trees as they are on the ground, are

A gray squirrel in its favorite pose on a tree trunk

described as "tree squirrels." One of them is the gray squirrel, wherever he is found—in the West, the Middle West, the East, and all through the South. He is larger than the red squirrel, and does not run through the trees with such speed. But he is well acquainted with them and with the electric wires around his home, along which he runs rapidly and surely.

He may be mostly gray, with white underneath, or he may have a tinge of rusty red on the head, sides, and legs. Or he may be very dark—even black. His coloring varies from one part of the country to another, but if the squirrel takes readily to the trees and is just a little larger than the red, you can be fairly certain that he is a gray squirrel.

Fox squirrel, an endangered species

FOX SQUIRREL

Another tree squirrel is the fox squirrel, which also comes in different "color schemes." In the South, a fox squirrel's head is likely to be very dark; in the central and northern states he may be almost as red as the red squirrel. In the Central Atlantic coast states fox squirrels are a beautiful silver gray. They are larger than gray squirrels. They climb trees easily and run along the stronger branches, but they do not leap swiftly from one tree to the next, and they are almost silent, without the chatter of the red or the somewhat quieter gray. They make the same kinds of nests and eat the same kinds of food.

Two kinds of fox squirrels seem to be threatened with extinction, occurring now in small numbers in a very limited area. One is the Delmarva Peninsula fox squirrel, which lives in only a few counties in Maryland, centering around Dorchester County. The other is the Everglades fox squirrel that is found only in Big Cypress Swamp and nearby pinelands of southwestern Florida, in Collier and northwestern Monroe counties. In both of these the loss of numbers is laid to the cutting down of pine trees suitable for nesting and bulldozing the area.

TASSEL-EARED SQUIRRELS

In the Southwest live two kinds of tree squirrels that have longer ears than other squirrels. The ears are pointed at the tips, and tufts of hair, an inch long or more, grow from the tips, making the ears seem even larger. The tufts are like tassels and give both squirrels a common name, the "tassel-eared squirrels."

One of these, the Kaibab squirrel, lives only on the North Rim of the Grand Canyon in Arizona. He is America's most beautiful squirrel—a big fellow with a big, fluffy, pure white tail that he carries like the proud banner of a knight. His body is mostly gray, tinged with red on the head and back, and his feet are white. He has always been few in numbers, because he lives only in a small area. Today he is very rare, and can easily become extinct. So he is protected by state and national laws.

The other "tassel-ear" is the Kaibab's close cousin, the Abert squirrel. He lives on the South Rim of the Grand Canyon, eastward into much of New Mexico, and northward into Colorado and Utah. He is much the same general color as the Kaibab, but is usually white underneath. His tail is dark above and white beneath, and it, too, is a handsome banner.

Kaibab squirrel

THE FLYING SQUIRREL

One kind of tree squirrel is different from the others in several important ways. He is the flying squirrel, and he lives in nearly all the United States and in most of Canada.

Flying squirrels have furry skin reaching from front foot to back foot. When the squirrel is sitting up or running, this skin lies in folds against his sides and is not noticeable. But, high in a tree, he can spread his legs and make a little kite or parachute of himself. Down he comes through the air, in a long, smooth glide. He is not really flying, but gliding, and can control his glide to land lightly on the ground or on a lower place on another tree. He can glide down—but not up. To go back up the tree, he climbs it just as any other squirrel climbs.

This is the only squirrel that regularly sleeps in the daytime and comes out at night. So we see very little of him, and that is too bad. He is a beautiful little animal, with the silkiest, softest fur imaginable, and huge soft, dark eyes. He is dark on back and head, white underneath. He is smaller than a gray squirrel and often smaller than a red squirrel, especially if he lives in the South.

THE CHIPMUNKS

Chipmunks are the smallest of all the squirrel cousins; they are about half tree squirrel and half ground squirrel. They don't go leaping through the trees like the red squirrels, but they are good climbers. They often go into trees for food and to get away from enemies. But just as often they run into a pile of rocks or into a hole in the ground.

A chipmunk builds his nest in the ground. He digs a long tunnel, straight down for a few inches, then slanting down to about three feet below the surface. He hollows out several rooms off the main tunnel, for use as a nursery, a storehouse, and a place to sleep. There is more than one entrance, and so, if a weasel or snake comes in the front, the chipmunk can go out the back.

He may use this same burrow all his life, lengthening the tunnel to as much as thirty feet. He winds it back and forth and adds side tunnels, and he adds more rooms. His burrow becomes a whole network of rooms and tunnels.

Chipmunks store food in their burrows, and in bad weather they do not have to go outdoors to eat. In the coldest weather they sleep the deep sleep of hibernation, going so sound asleep that people have handled them without waking them up.

There are so many kinds of chipmunks that it would take a whole book to tell about all of them. They are spread all over the United States and Canada. Take a picnic into the woods almost anywhere, and a friendly chipmunk is likely to come to it.

Most of the kinds are very much alike—reddish brown, with

black and white stripes along the body and face, and a brisk, longish tail that is often carried straight up when the chipmunk runs. The brilliance of his color varies, from one kind to another. In the Badlands of South Dakota, where visitors often feed the chipmunks from an overlook, they are the pale tans and pale reddish yellows of the rocks around them. In a forest they are likely to be much more strongly colored.

East of the Mississippi River, the chipmunk at your picnic is likely to be an eastern chipmunk. West of the river and northward through Canada, he may be a least chipmunk. Others live in smaller areas of the West—the Townsend in California, the yellow pine chipmunk in the Northwest, and the Colorado in Colorado and neighboring states.

THE GROUND SQUIRRELS

There are as many ground squirrels as there are tree squirrels —animals that make their homes in the ground and store their food there. There are many kinds of ground squirrels, and one or more kinds of them live in every part of the continent.

In some ways, ground squirrels are very much like chipmunks. One of them, the golden-mantled ground squirrel, even looks like a chipmunk. He is a little larger, and he has a splash of reddish gold across his head, throat, and shoulders that gives him his name. Because of the strong black and white stripes along his sides and his brisk and friendly ways, people watching him often think he is a chipmunk.

But look closely at his stripes. If they stop at his neck and do not go up across his face, you will know that you are watching a golden-mantled ground squirrel. Look for him in the West, especially in the mountain forests.

Although some of them can climb, ground squirrels spend most of their time on the ground. They tunnel into the ground

The golden-mantled ground squirrel, often mistaken for a chipmunk

to make their homes and storehouses, and run into a hole or tunnel if they are threatened. They eat seeds, roots, fruit, and almost anything else that tree squirrels eat, and they store food away that they eat in winter. Some of them hibernate in the coldest part of the winter.

From the Middle West all the way to the Pacific Coast, any spot out in the country is likely to have one or more kinds of ground squirrels. Look for them in meadows and fields, along streams, and on rocky hillsides and ridges, especially where there are evergreen trees.

The Franklin ground squirrel is common in the great plains and the Middle West, and the thirteen-lined ground squirrel, which has thirteen stripes, lives there, too. There are also many others.

Two friendly black-tailed prairie dogs

THE PRAIRIE DOG

Prairie dogs belong to the squirrel family. They dig tunnels near each other and make a prairie dog "town"—a colony. When they are outside their burrows, eating or taking a nap in the sun, one of them stands guard. He sits up tall—the picket pin again—and watches for danger. If he gives a sharp little bark, all of them instantly pop into their holes. So one moment you see a dozen prairie dogs, and the next moment there is nothing but sloping, bare ground with holes in it.

The native home of the prairie dogs is in the states of the great plains and the Rocky Mountains. These are states where cattle are raised, and hay and grain are grown to feed them. At one time, prairie dog towns often covered many acres and made them unfit for farming. Even in smaller "towns," cattle and horses often stepped into the holes. The result was usually a broken leg, and a valuable animal had to be killed.

So western ranchers decided to wipe out prairie dogs, and they have almost succeeded. Only a few prairie dog towns remain. One of these is in Wind Cave National Park in South Dakota, where a natural town has been preserved so that prairie dogs may never become extinct.

WOODCHUCKS AND MARMOTS

The woodchuck is another squirrel cousin, the largest of them all. Like the ground squirrel and the prairie dog, he is a great digger. He makes a long tunnel in the ground, and he likes to sit in the sun at the entrance to his tunnel. If an enemy startles him, he gives a loud whistle. Then, pop, into his tunnel he goes!

At the end of his tunnel is a comfortable burrow where he hibernates through the winter, after getting very fat during the summer and fall. There is usually more than one entrance to the burrow. Like many of the squirrels, he is clever at hiding the various entrances. As he digs, he may pile the dirt at the original entrance, then plug it up. He makes other entrances by coming up from below, into a bush or under a rock, where he

A marmot posing among his favorite rocks

leaves no telltale pile of dirt. The hole itself, well hidden, escapes notice.

Marmots are very much like woodchucks. They live high in the mountains, from the Rocky Mountains west to the Pacific. If you are driving on a mountain road and hear a loud whistle, a marmot is probably nearby. Watch for him in the rocks, and along the hillside. Often he sits on or partly under rocks at the very edge of the road.

MAKING GOOD USE

We have seen that members of the squirrel family make good use of their surroundings; they get food from the land and trees around them, and use them for protection from enemies and shelter from weather. Some of them, because there were never very many, are in danger of becoming extinct. Others, like the prairie dog, have crossed the interests of man and have come close to being exterminated. But they use their homes so effectively that many members of the squirrel family are almost certain to continue to live in great numbers.

You have read here about some, but not all the squirrel family relatives. Look for them and others near your home, and when you go for a trip. If you see a furry animal scuttling along a road through desert or prairie or meadow, or whisking up a tree—if it has a hairy tail—if it sits up like a picket pin to look you over—it very likely belongs to the squirrel family.

Chapter 5

THE HUNTED

We were visiting Custer State Park, in South Dakota, and a guide was taking us out to see some of the nocturnal wildlife in the park. It wasn't quite dark, and we could see a great bull elk with a magnificent spread of antlers on the crest of a high ridge, silhouetted against the sky.

Soon it was dark, and the night came alive around us. Frogs began peeping in a nearby marsh, and the mournful cry of a whippoorwill sounded from farther away. Our driver slackened the car and then came to a full stop as a mother and five baby skunks strung themselves out across the road in front of us. We watched in delighted amusement while they cleared the road, and then we went on.

Suddenly, as we rounded a curve, just ahead and a little to the right, we saw what looked like hundreds of tiny lights that resembled the lights of a city seen from an airplane.

"What is it?" we gasped.

Our guide stopped the car and turned his spotlight on the lights. There, on a small ridge, was a large herd of buffaloes, some lying down, some standing, but all with their heads turned toward our car. The lights were their eyes, reflecting our headlights.

The next morning we were in the guide's car as he led a caravan of autos through the park to see the buffalo herd by daylight. We found the herd near the entrance to Wind Cave

National Park, where two big bulls were facing each other, paw-
ing the earth and bellowing. As the cars braked to a stop, the
two bulls lunged at each other, banging their heads together
with a resounding crash that should have split their skulls, but
didn't! They did this several times, then turned around and
went off in opposite directions. In the meantime another bull
had gone off with the cow they were fighting over, and the rest
of the herd had continued quietly grazing, paying no attention
to the combatants.

Our guide had looked back to reassure anyone who might be
alarmed by the fighting, but, far from being frightened, the
tourists were all out of their cars, taking pictures. He urged
them to get back in their cars, for an angry buffalo bull is not
anything to get familiar with. He might take fright and run
away, but, on the other hand, he might lower his head and
charge right at you.

"Buffalo" is the popular name for this huge animal, but the

Buffaloes

scientific name is "bison." (The Latin name is *Bison bison.*) Often weighing a ton or more, the buffalo bull is an ungainly looking beast; but he is surprisingly light on his feet and can outrun a horse. Contributing to his ungainly appearance is a large hump that rises from his shoulders and slopes downward toward the hindquarters. An adult bull stands about six feet at the shoulder and measures ten to twelve feet from nose to tail. Dark brown, almost black in color, the hump and massive head are covered with long, shaggy hair, which extends down the front legs, looking somewhat like ragged pantalets.

The hair on his hindquarters is short and lighter in color. In the summer, when he sheds the thicker, warmer hair, the hindquarters become almost bare. Short, curving horns protrude above his eyes, which are so weak that he can see for only a short distance; but his hearing and sense of smell are unusually strong, seemingly to make up for the poor eyesight. The cow is a smaller version of the bull, weighing only 700 to 900 pounds and standing about five feet at the shoulder.

The calf, which is born in the spring, is a light yellowish red at first, but its color gradually darkens, until in a few months it is as dark as its parents. Within a few hours after birth, it is able to follow its mother and run with the herd. Its mother nurses it for nearly a year, until almost the time for a new calf to be born.

Buffaloes graze almost entirely on grass. They may eat the leaves on a low-growing bush occasionally, but they seldom browse on the taller shrubs and small trees as elk and deer do. The only use they have for trees is something to rub against to stop the itch caused by insects that make themselves at home in the short hair on the hind part of the animals. Another way to relieve themselves of these pests is to roll in the dust—mud is even better—often leaving deep depressions in the earth known as "wallows."

When the white man first came to America, there were millions of buffaloes grazing on the grasslands and on wooded areas, too many to be counted, but authorities estimate at least 60 million. They ranged from what is now Canada, down through the central plains to Mexico, and from Virginia in the East, across the plains and over the Rocky Mountains, to Wyoming and Montana. In the days of Daniel Boone, thousands of buffaloes made wide, smooth trails, likened to city streets, to Kentucky's salt licks. The greatest number, however, were on the central plains, where the grass was good and their only enemy was man. By 1830, all the buffaloes east of the Mississippi had been killed.

More gregarious than most other wild animals, buffaloes graze in small family groups, consisting of a bull and cow and several generations of her calves. The groups quickly merge into a large herd when they start to move to better grass, or when something happens to cause the herd to stampede. The bulls feed together peaceably, except during the breeding season, July and August, when they are apt to fight over the cows. When coyotes lurk along the edge of the herd or other dangers threaten, the calves are kept in the center of the herd, with the cows around them and the bulls on the outside.

IMPORTANCE TO THE INDIANS

Buffaloes were the way of life for the Plains Indians, for they provided food and shelter. However, before the white man introduced the horse and gun to America, hunting the great beast was both difficult and dangerous, and no part of the kill was wasted. Some of the meat was cooked and eaten immediately; some was cut into strips by the squaws and dried in the sun. This dried meat, called "jerky," could be carried on the hunt and eaten by the hunters. A favorite food of the Indians was "pemmican," which was dried meat pounded to a powder and mixed with buffalo fat. Berries and seeds were sometimes worked into it; then it was shaped into small cakes, which were edible for a long time.

Some of the skins were carefully scraped and tanned and the hair removed by the squaws and made into clothing or used to cover the tepees. When the hair was not removed, the skins, or "robes," were used for blankets and winter wraps. The bones had many uses: the large ones provided tools and weapons; the small ones, scrapers and other small tools. The ribs made runners for the dog sleds; the small, slender ones made good needles. The sinews provided thread and cord for bowstrings, and the horns were used for cups and ladles and ornaments for headdresses. These are only a few of the many ways the early Indians made use of the buffaloes they killed.

Before the coming of the white man, the Indians, using spear and bow and arrow, had to get close enough, on foot, to kill the buffaloes before the buffaloes killed them. When the Indians obtained horses, brought to America by the Spaniards, the going became easier. They could then ride alongside a herd, or around and around it, shooting with their bows and arrows as they went. Then each Indian claimed his kill by his particular arrows. But, even with horses, the Indians did not kill more than they needed.

SLAUGHTER BY THE WHITE MAN

Now came the white man. With his guns and his railroads, he did in thirty years what the Indians had not done in hundreds of years—he practically wiped out the huge herds of buffaloes. While the railroads were being built, buffaloes were killed in great numbers to feed the workmen. William Cody was given the name of Buffalo Bill because of the thousands he killed for the Union Pacific.

With the completion of the railroads, hides could be shipped east at little cost, and market hunters swarmed to the prairies. Often they took only the hides and the tongues, which were considered a delicacy. Sometimes in the winter they took the hump, but only a small part of the millions of pounds of meat could be marketed, and most of it was left where the animals had fallen, to spoil in the sun. It has been said that a person could walk for miles on the prairies and never step off of buffalo carcasses.

Sportsmen, shooting for "fun," added to the carnage. The railroads ran hunting excursions through herds of buffalo, and the so-called "sportsmen," from as far away as Europe, shot the buffaloes from car windows and let them lie where they fell.

The military encouraged the slaughter because the Indians would be easier to subdue when their principal food supply was gone. When concerned citizens finally demanded that something be done to save the buffalo, there were very few left to be saved.

EFFORTS TO SAVE THE BUFFALOES
FROM EXTINCTION

Before the end of the century the number was down to little more than 500, and the only truly wild ones were a herd of 21 in Yellowstone National Park, which was closed to hunting. In the vast wilderness of the park, they were able to roam free and safe from men and guns. The remainder were partly domesticated ones in parks and zoos and on the ranches of cattlemen, who had formed an association to save these great beasts from becoming extinct.

In 1907, 15 buffaloes were sent from the New York Zoological Park to the Wichita Mountains Wildlife Refuge, in the Wichita Mountains of Oklahoma, where 8,000 acres had been heavily fenced to receive them. Today that herd numbers 900, and each year buffaloes have to be sold to keep the herd at this number, which is all the refuge can support.

In 1908 President Theodore Roosevelt established the National Bison Range in western Montana, which protects about 400 buffaloes, along with other big-game animals, such as deer, elk, pronghorns, and bighorn sheep. The buffaloes in Yellowstone National Park now total 800, and there are nearly 400 in South Dakota's Wind Cave National Park. The greatest number of all is in adjoining Custer State Park, which has about 1,300. There are smaller numbers in other parks and refuges, and many ranches have small herds. All in all, the United States probably has about 25,000, and Canada has even more, with about 16,000 in Wood Buffalo National Park alone.

We shall never see great herds like those that roamed the plains during the last century, however, for there is no place for them. Their home is gone, covered by cities and towns, farms and ranches. Today the land that was once the habitat of the buffalo is now the habitat of people.

THE PRONGHORN

This fleet-footed creature is usually called a pronghorn antelope, but it is not really an antelope, although it resembles some of the antelope of the Old World. It came down to us from Miocene times, 10 to 25 million years ago. At that time, the time of grazing animals, it had numerous relatives, but our pronghorn is the only one of the family (Antilocapridae) to survive. All the other members have become extinct. It is truly a North American animal and is not found anywhere else.

The pronghorn is a handsome animal, with a rich tan coat and white underparts. The male has black markings on the face and on each side of the neck, but these markings are not as strong in the doe and may be lacking entirely. Both sexes have

Alert and watchful pronghorn

horns; on the doe they are about the length of her long, pointed ears, but they are much longer on the male. The tips of the horns are bent backward, and a prong, which gives the animal its name, juts forward on each horn. The horns are composed of a hard inner core and an outer sheath; this sheath is dropped every fall, and a new one grows over the core, which is not dropped.

The pronghorn is a small animal, weighing only 90 to 125 pounds; its best defenses are its eyes and its long, strong legs. The eyes are larger than those of animals several times its size, and it can see long distances with them, two miles or more. Its legs carry it over the ground at an unbelievable speed, thirty-five to forty miles an hour for several miles, with short spurts of sixty miles or more, and with long, horizontal leaps of fifteen to twenty feet.

When a pronghorn becomes frightened, it can erect the white patches of hair on the rump into two giant "powder puffs," which can be seen by other pronghorns a mile or more away. Each of them, in turn, erects the "powder puff," and then all dash away. When the hair is erected, a strong musk is released, the odor of which also serves as a warning to other pronghorns.

Pronghorns seem to enjoy a good run at top, or next to top, speed. Sometimes one or more of them will race an automobile on the highway for several miles and then turn abruptly and dash across the highway in front of the car and its startled occupants. They have also been known to race trains.

These graceful and interesting animals are also inquisitive. Hunters used to lure them within rifle range by waving pieces of cloth at them. They are too wary for that now, but they still can't resist standing immobile for an instant and taking a good look at whatever is threatening them before dashing away. One day, when we were driving in Custer State Park, two prong-

Pronghorns enjoying their favorite sport as they race at top speed across a Wyoming plain

horns in the road ahead of us came right up to the car; they then bobbed their heads for all the world as if they were making us a bow, and leaped to one side out of the road.

The fawns (also called kids) are born during May in the north and in late February and March in the south. After the first one, the doe usually has twins, which she leaves for several days lying on the ground a short distance apart and stays away from them except to feed them. The newborn fawns are reasonably safe, because they have little scent and blend so well with their surroundings that it is difficult to see them. When they are five days old, they can run faster than a man, and in about ten days they follow their mother and run with the herd.

Before the coming of the white man, there were millions of pronghorns in America; no one knows just how many, but estimates have been as high as 60 million. Like the buffalo, with which they sometimes mingled, they were crowded from much of their habitat and killed by hunters until their numbers were finally reduced to fewer than 20,000. Then the people woke up to the danger of their being exterminated, and all the states passed laws against killing them. Under protection, they came back very rapidly and now number about 300,000. Some of the states now have short open seasons on them.

They were able to make such a rapid comeback because they flourish in the barren lands of the West—plains, rolling hills, tablelands, and even deserts. They do not eat much grass, but browse mostly on shrubs and weeds and cacti and especially on sagebrush, which most other animals will not eat. So they are not a threat to cattle and other grazing animals. In the United States, the pronghorns live only in the western states. They are found in several western game refuges and national parks. Attempts to transplant them into the southeastern states and Wichita Mountains Wildlife Refuge have been unsuccessful so far.

SONORAN PRONGHORN THREATENED

The only pronghorn that is threatened with extinction is a subspecies called the Sonoran pronghorn, which is found in northwestern Sonora, Mexico, and the Cabeza Frieta Game Range and Organ Pipe National Monument in Arizona. A small, pale pronghorn, it is thought to be holding its own in Arizona, although there are fewer than 100, but it is believed to be declining in Mexico. It is, of course, protected in the Arizona refuges, though it sometimes wanders across the border into Mexico. The number in Mexico is unknown. There are perhaps as many as a thousand, which the Mexican government is taking steps to protect.

Chapter 6

HELP FOR
THE WHOOPERS

The great white birds rise majestically into the air, their satiny black wing tips flashing in the sun. Higher and higher they circle, their ringing, exultant cries echoing back to earth. Long after they are out of sight, their calls come back, carried on the morning breeze. It is April, and a whooping crane family has left its wintering grounds on the Texas coast to begin its 2,500-mile journey to its distant nesting grounds in northern Canada's Wood Buffalo National Park, a few hundred miles south of the Arctic Circle. Soon other cranes will follow until there are none left on the wintering grounds.

The story of the whooping crane, until 1968, is told in the authors' book, *Animals in Danger*, in the chapter "The Whoopers—Can We Save Them?" We still do not know the answer to that question, although two nations, the United States and Canada, are working together to try to save them.

This magnificent bird, standing nearly five feet in height, is the tallest native bird in North America. Its wings have a spread of seven feet or more. The adult crane is all white except for black wing tips, red skin on crown and cheeks, and stiff, hairlike black feathers across the cheeks. It has a long, sinuous neck and long, slim, black legs that can take it over the ground with great speed. The young whooper's feathers are brownish red in color, gradually turning white during its first year of life.

There probably never were great flocks of whooping cranes in North America. It has been estimated that possibly fewer than 1,500 were here, ranging from the Atlantic coast to Utah and from the Arctic coast to central Mexico, when the white man first came to this continent. Only a few wild ones are here now, but the number is increasing each year. Fifty-seven white-plumaged cranes left their wintering grounds on the Aransas National Wildlife Refuge in the spring of 1976, and 56 made the hazardous round trip of 5,000 miles, returning in the fall accompanied by 12 young ones. This is the largest number of young whoopers that has ever appeared at Aransas.

The whoopers migrate singly or in small groups, two or three together, and arrive at Aransas during late September, October, and November. Occasionally, as in 1975, a straggler or two will appear in December. In late March and April, they leave for their summer nesting sites, again in groups of two or three. No one knows why they take this long, arduous journey into the Canadian wilderness, but they are a very shy bird and apparently they want to establish their nests as far from man as they can.

For many years the location of the nesting sites was unknown, although diligent search was made for it. Finally, a forest fire in 1954 was the occasion for the disclosure of the nesting sites. A helicopter was returning from the fire in a northern section of Wood Buffalo National Park when the pilot and passenger saw two large white birds in a swampy area some miles south of the fire. Flying lower, they identified the birds as the long-sought whooping cranes. With them was their offspring, a small, long-legged, reddish-colored chick. Observations were immediately made from the air by the Canadian

Whooping cranes in flight

Wildlife Service, confirming that here, indeed, were the nesting sites of the elusive whooping cranes.

The next summer men from Canada and the United States made a careful survey of the area, by plane, helicopter, canoe, and on foot. They found a land remote from man and difficult of access, but ideal for the nesting needs of the whoopers. Each pair of whoopers requires a large territory in which to build its nest and rear its young. The area is black spruce swamp and muskeg, dotted with small lakes and ponds.

The nest is a large, flat mound made of marsh vegetation in shallow water. Two eggs are laid, but seldom, if ever, does more than one young whooper survive to accompany its parents to their wintering grounds. Nobody really knows what happens to that second egg. Sometimes it doesn't hatch, or the chick may be too weak to survive the dangers of its wilderness birthplace. The whooper chick has proved to be a pugnacious little fellow, and it has been suggested that the stronger one of the two may kill the weaker one.

A PLAN TO SAVE THE WHOOPERS

The Canadian Wildlife Service and the United States Fish and Wildlife Service got together and worked out a plan to save some of those lost eggs. First, U.S. scientists at the Patuxent Wildlife Research Center, at Laurel, Maryland, experimented with eggs taken from the nests of sandhill cranes, hatching the eggs in incubators and rearing the chicks by hand. When this proved effective, they decided, in the spring of 1967, that the time had come to put the whooping crane plan into action. United States and Canadian scientists flew by helicopter to the nesting area, where seven nests, six of them with two eggs each, had previously been located.

One egg was taken from each of the six nests, placed in a

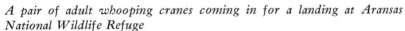

A pair of adult whooping cranes coming in for a landing at Aransas National Wildlife Refuge

portable incubator, and flown to the Patuxent Center. One egg
started to hatch on the way, but the chick died in the shell. The
other eggs hatched normally, but only three of the chicks survived to adulthood, growing into fine specimens, with well-
developed wing and body plumage. That fall nine young
whoopers accompanied their parents to Aransas.

On June 1, 1968, scientists again went to the nesting site and
took nine eggs and one chick from the whooping crane nests. A
second chick hatched from one of the eggs, and the chicks and
eggs were flown to Patuxent Center, where all the eggs hatched.
Later, at the Canadian nesting site, ten chicks hatched, eight
young cranes were seen later with their parents, but only six
chicks showed up that fall at their Aransas wintering site.

It was obvious that taking the eggs from the nests did not
lessen the number of cranes returning to Aransas each fall, and
so eggs were taken three additional summers, 1969, 1971, and
1974, and hatched in incubators at Patuxent Center. None were
taken in 1970, 1972, and 1973, but this did not result in more
young birds arriving at Aransas with their parents. In fact, in
1973 only two chicks appeared, while eight chicks had appeared in 1969 when eggs were taken. Five had appeared in
1971, and two in 1974.

In all, forty-nine eggs and one chick were taken from the
nesting site to the Patuxent Center. Thirty-seven chicks were
hatched, and of these, nineteen grew to adulthood, but one died
in the summer of 1975.

In the fall of 1972, thirteen adult-plumaged cranes failed to
return to Aransas from Canada, and no one knows what happened to them. They just dropped out of sight.

THE ARRIVAL OF DAWN

Whooping cranes are not known to reproduce before they are
at least five to seven years old, and the first pair mated at

Patuxent in 1975. When their first egg was laid, word was flashed to newspapers all over the country, but the egg was not fertile. The seven-year-old female laid two more eggs, both of which were fertile, but the chick died in the shell of the second egg. So, when a third precious egg was laid, it was taken from the mother and rushed to an incubator. Then the scientists at Patuxent—Dr. Ray Erickson, Assistant Director in charge of the threatened species research program; his behaviorist, Dr. Cameron Kepler; and other staff scientists—waited anxiously for the egg to hatch. In thirty days their patience was rewarded, as a whooper chick emerged from the egg, the first to be produced by the captive flock that had been hatched at Patuxent. Again word of its birth was flashed around the country.

The new chick was named Dawn, to signify the start of a new era in the fight to save the whooping cranes from extinction. Dawn was placed with a flock of turkey chicks, so that they could teach it how to eat and to get along with other chicks. This has been the policy with all the whooper chicks, because the larger number of turkeys helps to hold in check the pugnacity of the little whoopers, and also because the turkeys are more agile and can run faster than the whoopers and so avoid sustained attack and injury by the cranes. Unfortunately, little Dawn lived only two weeks, its death being due, probably, to a deformed leg.

A NEW PLAN

In May of 1975 a new plan was initiated. Fourteen eggs were taken from the nesting site in Canada, one from each nest where there were two. Instead of the eggs being taken to the Patuxent Center, however, they were taken to Grays Lake National Wildlife Refuge in southeastern Idaho and placed in the

Two-day-old whooper chick being handfed by a wildlife biologist. This chick was hatched in an incubator.

nests of greater sandhill cranes that make their home on the refuge during the nesting season. The sandhill crane eggs were removed from these nests.

Nine of the fourteen eggs at Grays Lake hatched, but one whooper was lost during a bad storm. The sandhill parents accepted the baby whoopers as their own, although they are somewhat different from the sandhill crane chicks. They are larger than the sandhill chicks, and their feet and bills are very dark, compared with the flesh-colored feet and bills of the sandhill cranes. Later, as the surface, or "contour," feathers of the whooping cranes begin to grow, they are a rufous-reddish color, contrasted with the paler rusty-brownish color of the sandhill cranes' feathers. Still later, at about four months, the contour feathers have emerged further to reveal white bases, giving a beautiful red and white mottled effect. The sandhill crane feathers at this age have gray bases; so the effect is a mottled rusty and gray. The whoopers retain their reddish color long after the sandhills have lost most of their rusty color.

Four young whoopers survived the summer at Grays Lake and the 800-mile journey to the sandhills' wintering grounds in New Mexico. One family settled on the Bosque del Apache National Wildlife Refuge on the Rio Grande, 85 miles south of

Albuquerque, and the others spread out through the central Rio Grande valley.

A Fish and Wildlife Service scientist drove south following the whoopers and their foster parents to keep track of their migration to New Mexico. This was the first migration of this sort to be closely monitored by man.

In the spring of 1976, the four young whoopers did not return to Grays Lake with their foster parents. Three of them were sighted at three widely separated locations in Idaho, Montana, and Utah, where they were doing well. The whereabouts, or fate, of the fourth one was unknown. In the fall these whoopers joined flocks of sandhill cranes migrating to the Rio Grande valley. To the delight of onlookers, one of the whoopers, with a flock of 19 sandhill cranes, stopped to rest and feed in a marsh in Ouray County, Colorado, near the town of Ridgway.

In May, 1976, fifteen eggs were taken from nests in the

This unusual grouping of adult whooping cranes and sandhill cranes shows the difference in size between the two species, as well as the darker coloring of the sandhills.

Canadian nesting site and placed under sandhill cranes at Grays Lake. Eleven of the eggs hatched, and four little whoopers survived unfavorable weather and predators to start the trip south to New Mexico with their foster parents. However, one young whooper was killed when it flew into a fence in Colorado's San Luis Valley.

As the nine pairs of captive whooping cranes at Patuxent mature and produce eggs, the plan now is to send these eggs to Grays Lake and place them under the wild sandhill cranes. This will avoid having to devise some plan to teach captive cranes to take care of themselves in the wild, a problem which appears to be almost unsurmountable. By this method it is hoped to create wild flocks that will follow much shorter migration routes than the present wild flock does. It is possible that as many as twenty-five eggs will be received from Wood Buffalo Park and Patuxent in May of 1977.

TROUBLE FOR THE WHOOPERS

In April, 1975, some of the whoopers that left Aransas so joyously for their nesting site in Canada ran into trouble on the way. Because of unseasonable snow and freezing rain, the flyway which thousands upon thousands of ducks and geese follow on their way north became blocked, and the waterfowl crowded into every available pond and lake and slough in southern Nebraska. The result was unprecedented overcrowding, and avian cholera, a deadly disease, broke out at the Sacramento Game Management Area near Kearney, and at several locations nearby. More than 15,000 ducks and geese died within a week.

This is an area where some of the whoopers often stopped to rest and feed, and state and federal wildlife officials watched for them apprehensively. Sure enough, nine unsuspecting

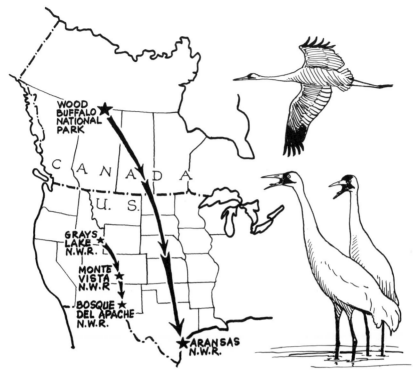

The long route of the wild whooping cranes across the United States and Canada and the shorter route of the sandhill cranes between Grays Lake National Wildlife Refuge in Idaho and Bosque Del Apache National Wildlife Refuge in New Mexico

whoopers showed up and settled into an infected marsh before they could be stopped. Everyone who could—state and federal wildlife officials, farmers and other people living in the vicinity, and the crews that were clearing away the dead waterfowl—took part in the attempt to drive away the whoopers. Low-flying aircraft was used and firecrackers and other ground noise-makers. It took thirty-six hours of frantic work to flush out the big birds and get them started to a safer place, but the efforts to save them were successful. Avian cholera kills infected birds within a short time; since all the forty-nine cranes that had left the Aransas Refuge in the spring returned there in the fall, it is apparent that all the whoopers had reached their destination in Canada safely.

BOMBING RANGE ABANDONED

Early in 1975 the Air Force announced that it was abandoning its bombing range on Matagorda Island, next door to the Aransas Refuge. This bombing range had long been a subject of controversy because it was feared that the sound of bombing and the low-flying planes frightened the whoopers and were contributing to their dwindling numbers. The Fish and Wildlife Service hopes to add the island to the Aransas Refuge, but in any event, the whoopers will be able to feed in the inland estuarine areas, which are rich in crabs, shrimp, frogs, small fish, and other water animals that are the whoopers' favorite food.

The Fish and Wildlife Service has also worked out a plan to close temporarily to hunters areas where migratory birds feed whenever whooping cranes or other endangered species are approaching the areas.

It looks as if the whooping cranes may be saved—for a while, at least.

Chapter 7

MAN-MADE HAVENS

He fixed us with a solemn, round eye, almost seeming in a dignified way to glare at us. Did he think we were interfering with the appearance of his food? Food was all he seemed to have on his young mind.

He was a baby brown pelican, named "Pax" for peace, hatched three days ago, and he was virtually unique—a living pelican hatched in captivity. He was a gawky, skin-and-bones, purplish pink caricature of his parents. Almost the only motion he displayed was a bobbing of his head while the tip of his bill moved in search of food that his parents had regurgitated. In a very few days he would know enough to ram his head deep into the pouch of one of his parents in search of the regurgitated food.

As he grew older, white down, then brown feathers would cover him. At twelve to fifteen weeks he would be mature enough to fly. And at fifteen weeks, he did. For several days he exercised inside the sanctuary, flapping his wings at a great rate and flying briefly from one spot to another. Then one evening at sunset, after he was fed, he flew to the fence and kept going, straight north along the shore. Pax was now on his own.

He did not come back, although it was hoped he would be a "fly-in"—a bird that comes back for the night. But Pax was truly a wild bird—one raised by his parents without human help—and he had returned to the wild.

After Pax's departure, the parents began to make preparations for nesting again.

THE SUNCOAST SEABIRD SANCTUARY

"Captivity" was hardly the condition in which Pax's parents lived and into which he had been thrust on hatching. They were more like the recuperating guests of a top-grade nursing home or hospital; they were guests at the Suncoast Seabird Sanctuary at Indian Rocks Beach, near St. Petersburg, Florida.

Alexis, the mother, had been taken to the sanctuary when she was picked up, at Johns Pass nearby, a very young, wet, bedraggled bird and unable to fly, and she has been at the sanctuary ever since. Mature now, she could fly—but she chose the way of life of the sanctuary. Her mate, Salty, had been taken there with serious injuries to his wing from a fish hook. He would never fly again.

The two of them had made their nest of sticks on the ground, at the back of the sanctuary against a board wall. Salty had done most of the incubation of the egg, and was the one most frequently crouched over the chick. He couldn't get away from the sanctuary, and it seemed to be Alexis's idea that the best way to keep him out of trouble was to keep him busy.

There are some 150 pelicans at the sanctuary, all marked with a red or a yellow tag. They are brought in at the rate of several a day—among about 15 sick or injured birds a day, in all. The red tag gives the information that the bird can fly again, the yellow that it is permanently crippled. When a bird that can fly is released into the compound, it usually goes its way. Sometimes it comes back to spend the night, a fly-in. There are several pelican fly-ins.

The birds were not at all disturbed by us (and a lot of other

The newly hatched baby pelican, Pax, is sheltered by his mother, Alexis, at the Suncoast Seabird Sanctuary in Florida.

people) watching them. "This is my home," they seemed to say. "Nothing bad can happen to us here."

The sanctuary is owned and operated by Ralph T. Heath, Jr., assisted by two full-time helpers and a number of volunteers. Each day they care for, feed, and clean cages for 600 birds of about fifty species. Doctoring is done by veterinarians, called in from the outside, and carefully regulated. Besides pelicans, the "sea birds" include American and snowy egrets, great blue herons, cormorants, purple gallinules, ibises, sandpipers, terns, gulls of several kinds, and various others. A little green heron was brought in with a broken jaw, and part of its care consisted of forced feeding without moving the splints on the jaw—a procedure which the workers successfully mastered.

Land birds at the sanctuary include turkey vultures, red-tailed hawks, screech owls, a great horned owl, a burrowing owl, a chuck-wills-widow, myrtle warblers, and various others. Proof of public interest is more than 100 phone calls daily from people asking about certain birds, or for information about taking care of one. People flock to the sanctuary in the evening to see the fly-ins arrive for evening feeding. Mostly pelicans and egrets, they present a rare sight as they circle and drop into the compound.

AN ADOPT-A-BIRD PROGRAM

A pound of fish a day is fed to each pelican. This does not more than start the feeding operation for the whole sanctuary, which runs between $700 and $800 a month. Financial help comes to the sanctuary only through donations, and a unique method for them has recently been put into action. This is the "Adopt-A-Bird" program, in which an individual, or a school, or an organization, picks out a bird to adopt and pays a given amount a month for its support. Several hundred of the sanctuary's birds have already been adopted.

Different kinds of birds cost different amounts. A pelican

Ralph Heath of the Suncoast Seabird Sanctuary feeds some of the pelicans that were brought to the sanctuary with fishhooks embedded in them, or with other injuries.

Fluffy, a great horned owl, was brought to the sanctuary for treatment of injuries.

costs seven dollars a month, an egret five, an eagle thirty. Mr. Heath remarks dryly that this is one case where the eagle's bill is bigger than the pelican's.

TWO YOUNG EAGLES

Notable among the sanctuary's land birds are two young bald eagles. One was picked up at the foot of the tree that held its nest. It was too young to fly, and evidently in trouble because it was only skin and bones. Examination at the sanctuary disclosed a fish bone in its throat, which a veterinarian removed, and the bird immediately began to eat and to gain weight.

But its troubles were still not over. The veterinarian discovered that the bone in the first joint of a wing was broken. So this had to be set in splints, and the eagle kept at the sanctuary until the bone healed. By the time it could be released, its family, with its brother eaglet, had flown north. Would the young eagle follow them on release? No one would know, unless a report came in from the U.S. Fish and Wildlife band on its leg.

The second young eagle was fished by two teenagers out of a small lake, where it had fallen, and was taken to the sanctuary. Both joints on its left wing had been damaged in some way prior to its fall. The bird was taken to a veterinarian hospital, where the wing was set with all the safeguards and protection of a modern hospital. It will be cared for by the sanctuary until it can fly again. It is hoped that both of the young eagles can eventually be returned to the wild, able to fly and to get food for themselves. But this will take at least a year's stay at the sanctuary, for training and conditioning.

EAGLES STRUGGLING TO SURVIVE

The young eagles, and all young eagles, are of special value, because the species is still dwindling. The southern race of bald eagles is listed in the latest of the United States Fish and Wildlife Service's *Threatened Wildlife of the United States* and noted as generally decreasing throughout its range, which is in nearly all parts of the southern half of the United States.

The northern race of bald eagles is also still on the wane, although it was not listed as threatened in the Fish and Wildlife's publication. It is undergoing the same hazards as the southern eagles—disturbance by people of its nesting areas, illegal shooting, loss of nest trees, and reduced hatching because

Doctors removed a bone from the throat of this young bald eagle named Centurion, after he fell from his nest and was taken to the sanctuary. Later it was found that a wing bone was broken.

of pesticides taken in with food. Only the bald eagles of Alaska seem to be holding their own.

In Minnesota, especially on the national forests, the number of successful bald eagle nests seems to be encouraging, with more than 100 nests where healthy eggs were laid and more than 100 young eagles produced. This contrasts with 30 to 40 pairs of eagles in the state of Maine, producing only seven eaglets. This low number was apparently brought about by the presence of insecticides in the eagles' food, causing the egg-shells to become too thin for incubation or the eggs themselves to become addled. So the U.S. Fish and Wildlife Service, in conjunction with several other conservation agencies, tried

Centennial, a young bald eagle fished from a lake and taken to the sanctuary with a broken wing

transplanting healthy eggs from Minnesota to Maine nests; from two eggs successfully moved, two young eaglets were hatched, and the technique of transplanting was demonstrated as sound for future development.

In Wisconsin and the Upper Peninsula of Michigan the number of bald eagles on the U.S. forests is encouraging. In the Sucker Lake area of Ottawa National Forest is what the foresters call a "bald eagle pad." It is an area where immature eagles—between the time they leave the nest and the time they mate, a period of about six years—club together and eat and fly and grow strong.

Alaska is still a bald eagle stronghold, with a stable population of between 30,000 and 55,000 birds.

Great effort is being made to protect the eagles. Both bald and golden eagles are protected by federal law, and severe penalties are imposed for violations. Nesting sites are protected in many localities, in such areas as national wildlife sanctuaries and sanctuaries of the National Audubon Society. In both, trees exist where nesting occurs or in promising sites. These are not disturbed by visitors, timbering, or other activities.

The Florida Audubon Society alone has agreements with a growing number of landowners for eagle sanctuaries; more than 2,300,000 acres have been put under contract. Pesticides are under study by federal agencies to determine their effect on eagle reproduction and the necessity for greater limitation of pollutants, as well as for methods for production and care of young in various localities.

FIRST IN A GREAT SYSTEM

So much is said these days about the need for man to protect the homes of animals, that it might give us the impression that this is a new idea. Actually, an organized attempt in that direction was started nearly a hundred years ago.

In 1903, a man named Kroegel lived in the little town of Sebastian, on the Atlantic coast of Florida. Two miles offshore was a tiny island of about three acres. Kroegel had watched this island for years—had seen thousands of pelicans come there every year to make nests, lay eggs, and raise their young. He was anxious to protect this breeding ground of the pelicans and other seabirds as well.

The island was very vulnerable. It was in plain sight of land, and of pleasure boats that ran up and down the coast. People in the boats shot the pelicans for sport. Fishermen decided the birds were getting too many fish, and a crew landed on the island and clubbed many of the young birds to death.

For a long time, Kroegel dashed out whenever he heard a boat nearing the island, and warned people away with his shotgun. But he had no real authority, and his protection fell far short of what was needed.

But in 1903 "Pelican Island" was made a United States bird refuge by President Theodore Roosevelt; it was the first of what was to become nearly 360 United States wildlife refuges, spread far and wide into all parts of the country. Kroegel was put in charge of Pelican Island, and so he was the first refuge warden.

The island refuge was expanded to take in other nearby islands and the sea between them. Today the refuge has 756 acres sheltering thousands of breeding pelicans. Of the 25,000 birds that come to the islands to make their nests, there are great blue herons, snowy and American egrets, little blue herons, green herons, anhingas, roseate spoonbills, and many others. Some 250 white pelicans come here to spend the winter from their breeding grounds in the West.

So began what is probably the greatest effort the world has known for protecting wildlife in its natural surroundings. Three hundred and fifty-six national wildlife refuges with more than 30 million acres are scattered in all fifty states. Hundreds of thousands of acres are devoted to waterfowl in their natural phases of migration and breeding. Miles of ditching and dikes through endless prairie acres have helped to offset earlier drainage, holding the water so necessary to the ducks and geese that use it along their migration routes. Every year, Tule Lake National Wildlife Refuge in California has the greatest concentration of waterfowl on the continent.

A goose pictured in flight is the symbol of the refuges. A sign showing it in a prominent spot appears on each refuge.

Special projects have reached out to species that are in immediate danger of extinction. So mention of the Aransas

"*The Sign of the Flying Goose*" *is the official marker of the United States national wildlife refuges.*

Wildlife Refuge in Texas calls immediately to mind the endangered whooping cranes; two other refuges are involved, the Bosque del Apache Refuge in New Mexico and Grays Lake Refuge in Idaho, where attempts are being made to start additional colonies of whoopers.

In the Aleutian Islands National Wildlife Refuge in Alaska, important work is progressing in the study and protection of the sea otter, of which the southern subspecies off the California coast, is threatened. While these animals are gaining, they still number only a few more than 1,000.

Red Rock Lakes Refuge is the official home of the trumpeter swan, whose successful comeback has allowed it to be transplanted to fifteen other refuges in western states.

Loxahatchee Refuge in Florida centers the range of the Florida Everglades kite, of which there are only about fifty birds. It is sometimes seen in areas bordering the refuge, where nesting sites are protected from disturbance by visitors. Propagation of the apple snail, its only food, is being succesfully encouraged in the refuge and elsewhere.

There are many others—the Key Deer Refuge in Florida protects the tiny Key deer, which is on the increase there, but still probably is only somewhere around 600. It lives on Big Pine Key and other islands in the vicinity that have fresh water and pine and hardwood woodlands. Acquiring of other islands with this special kind of home, as part of the refuge, may safeguard enough of the Key deer homeland to protect them from becoming extinct.

The Hawaiian Island Refuge protects the Hawaiian monk seal, the Laysan finch, and others; the Wichita Mountains Refuge in Oklahoma protects the black-tailed prairie dog in a colony on the increase (one of several).

In the areas where prairie dog colonies are protected (including Wind Cave National Park in South Dakota) an occasional glimpse is seen of the black-footed ferret, probably our

Key deer in typical home surroundings on National Key Deer Refuge on Big Pine Key in Florida. This buck measured only 22 inches at the shoulder.

The black-footed ferret

rarest mammal. It depends on the prairie dog for food. Some authorities think it may not be as rare as reported, but, extremely shy, is just not seen very often. One observer reports watching a prairie dog town for two weeks and finally catching a glimpse of a ferret.

More than thirty-five threatened species are aided in some way, at one season or another, by the wildlife refuges. At the Bear River Migratory Bird Refuge, for example, many thousands of ducks and geese are nesting, and there are white pelicans, ibises, herons and egrets, and many others. Various other refuges that do not bear this exact title come within the system. The National Bison Range in Montana not only protects buffaloes but a big population of pronghorns, bighorn sheep, elk, and deer.

ESTUARINE SANCTUARIES

Somewhat similar to the national wildlife refuges is a group of estuarine sanctuaries developing in coastal zone management. Most recent of these is Sapelo Island Estuary on the Duplin River, where about 5,800 acres of marshland and high ground on Sapelo Island, just off the southeast coast of Georgia, will be maintained in their natural state. Here the tidal salt

Many different kinds of coral and countless other forms of animal life are preserved in their natural state in this estuarine sanctuary off Key Largo, Florida.

marsh and the Duplin River and numerous tidal creeks are typical of estuaries up and down the coast, and will be used for study of such an environment for decisions in coastal management. There are a great many estuaries scattered along America's coasts, and money is provided to the states by the federal government for protecting the natural environment of the estuary and the living things within it.

Other estuarine sanctuaries include 4,300 acres in the South Slough Sanctuary on the Pacific Ocean near Coos Bay, Oregon. Further proposals include 100 square miles of living coral reef off Key Largo, Florida.

THE NATIONAL PARKS

Other United States-owned havens offer sanctuary to wildlife. These are the national parks. Some of these are very important to the preservation of decreasing wildlife.

At one time, Yellowstone National Park had the only trumpeter swans in existence. Biologists worked with them in the park for years before they established the number sufficient to survive transferral to other areas. Yellowstone also is famous for the herds of elk and buffalo that are protected there.

The North Rim of Grand Canyon National Park has one of the finest herds of deer on the continent; and the North Rim— the only area where it lives—has the very rare Kaibab squirrel. There are about 4,000 of the squirrels living on the North Rim; but their homeland is changing because of fire protection in the area. They seem to prefer a habitat of burned-over forest.

On the South Rim is a cousin of the Kaibab, the Abert squirrel; the Abert also lives south of the canyon and eastward and northward into New Mexico and Colorado. These two are at home and find much of their food in the yellow pines; they eat the cambium layer of the bark.

Certain national parks are the home of the endangered mountain lion.

The dwindling number of mountain lions find sanctuary in the national parks. Individuals are sometimes seen on the South Rim of the Grand Canyon and in the parks of the Rocky Mountains. In Florida under a different name, the Florida panther is protected in Everglades National Park, where its number may have dropped to 100 or less. It has been known in recent years to travel north from the park to Georgia.

Closely related, if not the same animal, the Eastern cougar was thought to be extinct, but in recent years sightings have been reported in the far Northeast; one authority reported about twenty-four sightings in New Brunswick, Canada.

Timber wolves also seem to be making their last stand in national parks. An established band lives on Isle Royale National Park, in Lake Superior; a greater number roam the wilderness of the new Voyageurs' National Park in Minnesota, in the Boundary Waters Canoe Area. Altogether there may be close to 2,000. A similar animal, the northern Rocky Mountain wolf, once thought to be extinct, has been reported recently in Yellowstone and Glacier national parks.

Voyageurs' National Park also reports protection of a promising number of bald eagles and their nests, and, as well, of the

The timber wolf finds safety in Voyageurs' National Park, Isle Royale National Park, and several others.

American osprey. The latter, while not endangered, is another bird that is losing a great deal of its natural habitat. And Voyageurs' controls and protects another species, a very uncommon creature—the lake sturgeon. Here the fish is given protection to allow it to reach maturity, which takes about twenty years. If it is "harvested" at an earlier age, its numbers are very rapidly reduced.

Probably the best-known park in the protection of wildlife is Everglades National Park in Florida, where at least two major battles for preservation of a species have centered. One of these concerned the alligator, whose present range runs from North Carolina southward along the coast of Florida, along the Gulf Coast to Texas, and into Arkansas and Oklahoma. Much of the poaching of big alligators took place in the Everglades park, where, before federal laws protected them, it fell to the park to patrol against poaching. Action by the park rangers stopped much of it, and countless alligators were saved.

Everglades National Park is, too, the scene where the battle to protect the plume birds was fought and won with the preservation of several species. American egrets and snowy egrets— the "golden slipper bird" because of its yellow feet—and the roseate spoonbill all produce beautiful plumes, and all were ruthlessly slaughtered to secure their plumes for women's millinery. But before the slaughter was complete, laws were passed against killing them, and Everglades National Park was formed. Here the great nursery rookeries in the tops of the mangroves on islands and along the shores of the mainland are watchfully protected. In the breeding season, when the plumes are most beautiful, thousands of nests are made in safety here, and thousands of young are raised to the stage at which they can leave the nest.

Many other kinds of birds and animals are protected here, too, living most of their lives within the park boundaries. There

This rookery is "home" to many different kinds of birds.

are great white herons, great blue herons, bitterns, owls, and many other birds, as well as bobcats, raccoons, and other mammals.

FOREST HAVENS

Many birds and other animals are at home in the United States national forests and in the various state forests. As we noted earlier, bald eagles are protected from disturbance in the national forests of the Middle West.

Special effort is being made on Superior National Forest in Minnesota to provide food for timber wolves—one line of work among many to encourage species that are dwindling. Here the foresters are cutting timber in order to start new young trees on which deer and moose feed, so that their numbers stay up enough to give the timber wolf enough meat to carry it around the year. This attention to the food chain of threatened animals is typical of the activities of the Forest Service wherever there are national forests.

Probably the most famous birds to find protection in the national forests are the California condors, which have most of their nesting and roosting areas in the Los Padres National Forest of southern California. There are two sanctuaries, the Sespe Wildlife Area and the Sisquoc Condor Sanctuary, set aside specifically to protect the condors from being disturbed.

The condors, of which there are only between forty and sixty, are easily disturbed, and will, with patience and persistence, watch anyone trying to watch them. If people come too close or there is too much noise in the vicinity, the birds will desert their nests. The critical acreages have been threatened by oil drillers, road builders, and dam builders, but so far these interests have been turned away. Visitors are not allowed to approach the nests. The Forest Service combines with several other agencies in protection of the sanctuaries.

Careful management and vigilance are also the watchwords for a special area set aside in Michigan, in the Huron National Forest and in an area of the Michigan State Forest—more than 4,000 acres in all. These are jack-pine acres, and they have the threatened Kirtland's warbler nesting at the foot of the young trees—the only situation in which it will nest. "Operation Popcone," in which controlled acres are burned over before the warblers arrive from their winter range in the Bahamas, was successful in making the jack-pine cones open and plant new trees. But the population of birds dropped alarmingly in recent years to fewer than 300 pairs, thought to be largely because of the presence of cowbirds. The cowbirds laid their eggs in the warblers' nests and caused the warblers' young to be diminished. Study of the cowbirds resulted in ways of controlling them and stopped the downward trend in the warblers' population; but there are still fewer than 300 pairs.

More than fifty of the rare or endangered species of wildlife are to be seen in or near the national forests. The bald eagle, the Kaibab squirrel, the Tule elk, the bighorn sheep, mountain lions. wolves grizzly bears, and many others take national forest land into their ranges. Wherever these animals are found, the Forest Service gives special emphasis to management of their habitat, so that their living conditions can be improved and they can be guarded from changes that might contribute to extinction of their species.

HAVENS PROTECTED BY ORGANIZATIONS

We walked along the boardwalk that led into 11,000 acres of freshwater swamp—Corkscrew Swamp. Below us, very near, was a young wood stork wading in the water, and very near him was an alligator. The alligator didn't seem to be paying much attention to him, but every now and then the alligator would move a little, and would manage the move so that it brought him a little closer to the wood stork. The wood stork didn't seem to be paying much attention, either, but whenever the alligator came closer to him, he would move a little farther away. Apparently he liked to live dangerously, for as long as we stayed there, the two of them kept up this game, the wood stork teasing the alligator, and the alligator trying to work close enough to strike at him.

The boardwalk was a mile long and wound through the swamp to show us many sights. Water plants stretched along the water's surface, and bald cypress trees grew in the water, the biggest stand of bald cypress in the country. More than 100 feet up, their crowns held the nests of a huge colony of wood storks, birds that laid their eggs and hatched their young and fed their young birds without being disturbed.

This is a sanctuary owned by the National Audubon Society; its boardwalk allows visitors to see some of the swamp without disturbing the wildlife in it. Its wood stork nursery is one of the largest in the country and usually produces a big crop of young birds to assure the wood stork a thriving new generation. The weather is sometimes a hazard, but if cold weather spells disaster for the young birds, the parents have been known to produce a second crop.

Corkscrew lies in the heart of Florida, to the north of Everglades National Park. It is threatened by drainage projects in its vicinity, but the Audubon Society and other agencies have been active in protection of the sanctuary and have been able to induce developers to plug up some of the drainage canals where they are useless and destructive to the swamp. The massive drainage programs throughout southern Florida seem to have increased the severity of the annual periods of drought and flood, with an adverse effect on the food supply of the wood storks. The storks in the sanctuary as well as those outside it have been affected, and the whole population is undergoing a gradual decrease.

Corkscrew Swamp is only one of more than forty sanctuaries operated by the Audubon Society, which devotes some 20 percent of its annual income to this kind of project. Corkscrew Swamp is unique among the Society's sanctuary areas, in that one need not make special arrangements to visit. To protect the others from too many visitors and the disturbance that they might make, the Society requires that permission to visit be obtained. Permission may be obtained by writing to Sanctuary Department Direction, Miles Wildlife Sanctuary, West Cornwall Road, Sharon, Connecticut 06069.

A boardwalk through Corkscrew Swamp lets visitors view its birds and other animals without disturbing them.

Some of the sanctuaries are the Vingt-et-un Wildlife Sanctuary near Smith Point, Texas, which protects the largest colony of roseate spoonbills—400 to 500 pairs; the Green Island Wildlife Sanctuary near Rio Hondo, Texas, which has the largest colony of reddish egrets—about 1,500 pairs; and Lake Okeechobee Wildlife Sanctuary near Okeechobee, Florida, where there are approximately 25 pairs of Everglade kites. Other organizations maintain somewhat similar sanctuaries, as do many counties and states, protecting all kinds of wildlife.

MANY AGENCIES CONCERNED

These are only some of the agencies that are employing a wide range of protection for threatened wildlife. Every state has its research units, for example, and its agencies for securing legislation in protection of species that are or might become endangered. All across the land, colleges and universities and organizations act in conjunction with government agencies.

The lesser prairie chicken is a good example of the concern of several cooperating states—Kansas, Colorado, New Mexico, Texas, and Oklahoma. Facts are hard to come by concerning the prairie chicken, because of a baffling great fluctuation in numbers—as much as 2,000 to 30,000 birds—but its overall status seems to be decreasing with great vulnerability as it faces the future. All agencies within these states are cooperating in protecting it, with no hunting and with brush and fire control. The various states have acquired big areas where there is a suitable grasslands habitat for this bird.

Of even more concern is Attwater's greater prairie chicken, which exists in small numbers in widely scattered areas in the Gulf coastal prairie of Texas. Here again sanctuary areas have been acquired by various agencies, visitors' activities are limited, and there is vigilant fire protection.

Attwater's greater prairie chicken

Actually, there is probably today no species on the wane that is not the subject of concern of several agencies—some species being the concern of many agencies. Yet sometimes the best that maximum effort and cooperation can do is not enough. The peregrine falcon, for example, once ranged over most of the United States and southern Canada. In recent years it has steadily lost ground, until it no longer nests east of the Rocky Mountains or in many former favorite areas of the West. In a recent year, only three young were produced from the fourteen nests reported in the southern Rocky Mountains.

Worst villain seems to be the presence of insecticides in the adults' food, causing the shells of the eggs to be too thin to survive the processes of incubation and hatching. Checking the shells has shown them to be 20 percent thinner than those of the falcons in Canada. Another danger comes from poachers who remove the young falcons from the nests and raise them in captivity, for use in falconry.

Adverse effects continue in spite of the combined efforts of many agencies. Special agencies in all the states where the falcons are known, teams from colleges such as Colorado State University, and teams from national organizations such as the World Wildlife Fund, National Audubon Society, and Defenders of Wildlife.

Peregrine falcon

Two recovery teams, one in the East and one in the Rocky Mountains, have been established by the U.S. Fish and Wildlife Service, made up of experts from all over the country. Their activities include releasing falcons near old nests, guarding active nests, and substituting healthy eggs laid in captivity for those known to have been damaged by insecticides in active nests.

On the East Coast, young falcons were recently released through the joint efforts of the U.S. Army, the U.S. Fish and Wildlife Service, Cornell University, and others. Cornell maintains facilities for breeding the birds in captivity, and this was the source of supply for the young birds. Releases were made on Carroll Island in Maryland; in upstate New York, near Cornell University; in the Catskills; and in Massachusetts.

Releases have been made at several sites in the Rocky Mountain states, and a western breeding facility has been established at a research site in Colorado. Falconers, raising young from falcons in captivity, have released many of them into the wild.

Nests in the wild have been placed under the strictest watch to prevent disturbing the birds. Morro Rock in California, for example, is a favorite nesting site for the falcons. It was raided for young birds, and later was disturbed by people going too near. So an electronic alarm system was installed, to keep visitors from penetrating to the nesting site. But still the number declines; the peregrine falcon is one of our very rarest birds.

The California least tern is a rare bird that nests along the Pacific coast of California, from South San Francisco Bay to southern Baja California. Its nest is only a shallow depression in the sand, where it lays two eggs. Its breeding grounds were destroyed to such an extent that it was reduced to fewer than 300 pairs. Several agencies moved to protect the tern, establishing sanctuaries where it made its nests. The United States Navy cooperated, too, protecting nesting at Seal Beach Naval Weapons Station. Several sanctuaries were established by the state, and one was set up at Camp Pendleton Marine Base. All of these are maintained so that this little bird can nest in safety, protected from dogs and the wheels of motor scooters and dune buggies. The future may hold nearby artificial islands dredged up offshore from protected areas, where the birds can nest.

It would seem that with so many people working together for the protection of threatened species, we should soon be able to say that no species is in danger. But stopping the downward trend is a long-time and complex problem—one that will demand for many years to come the activities for which man-made havens are established.

Chapter 8

MOUNTAIN AND CLIFF CLIMBERS

The great bighorn ram climbed steadily up the mountainside. Below him, in a bowl of mountains, was the little town of Ouray, Colorado. Above him was a giant cliff that towered almost straight up the mountainside. Behind him was a pack of dogs from the town that were running loose.

The ram was making for the cliff. Would the dogs catch him there—would he have to face them, his back to the stone wall?

When he reached the wall, he kept on climbing, along a little ledge that was so narrow it could not be seen from a little way off. As he climbed along it, the dogs followed, one at a time, until all five of them were on the little ledge, in single file.

On he went, not pausing at all as the little ledge became more and more narrow. Behind him came the dogs, barking and baying, eager to get close enough to him to pull him down.

But that was not to be. The little ledge finally became so narrow that it seemed to disappear altogether. But the ram's feet took him along it, higher and higher, where the dogs could not go. At last he reached the top of the cliff. He stood a moment and looked down at the dogs, strung out on the narrow ledge. Then he tossed his head and, turning, went to join some other bighorn sheep that were grazing in a flat little grassy area ahead.

116

Below, the dogs began to howl. They could not follow the ram up the cliff. They could not turn and go back down. They were stuck on the ledge, and all they could do was howl.

People in the town heard them, including their owners. The owners were very much embarrassed, because it was against the law for the dogs to be chasing the mountain sheep. But they could not leave the dogs to almost certain death on the face of the cliff. They had to climb up the mountainside and help them get down. Then they had to pay a fine because their dogs had been marauding the sheep.

The ram, of course, did not know this, but he knew he had stranded the dogs, and he went on quietly grazing. After awhile he led the mountain sheep to a still higher mountain meadow where there was more food for them.

ROCKY MOUNTAIN BIGHORNS

It is thought that there are about 12,000 Rocky Mountain bighorns living in the West, but an accurate census of them is very hard to make. Their favorite home is among the highest crags of the mountains near and above timberline, and they are constantly on guard and move away if an intruder appears.

They are scattered in small bands in the Rocky Mountains, from Canada into New Mexico. Some of the bands are in national parks, especially Yellowstone, Glacier, Teton, and Rocky Mountain National Park. There is a band in the National Bison Range north of Missoula, Montana. They live in the highest peaks and meadows in summer, and migrate to lower areas of less snow when winter comes.

A large-sized bighorn ram is perhaps seven feet long and stands as tall as three-and-a-half feet at the shoulder. He is a stocky, powerful animal, his strong muscles bulging in legs, thighs, and shoulders. He may weigh up to 300 pounds. He has heavy horns that sweep up and back, then forward again, in the beginning of a spiral. Along the outer edge, the horns have been measured as long as forty-eight inches, and they may be sixteen inches around at the base. They continue to grow throughout the ram's life. The ewes are a little smaller than the rams, and their horns are almost straight—sharp spikes six to eight inches long.

The bighorns have very fine sight and are able to see an enemy creeping up on them from some distance away. Their enemies include mountain lions and bobcats. An occasional lamb is taken by a golden eagle.

They eat the grass that grows at high elevations, the tiny annuals that come out in the spring, and low bushes. Here, again, they have an enemy—the domestic sheep that are brought by man to the mountains to graze, which compete for the already slim food supply. But their greatest enemy is disease and the parasites that live inside their bodies.

The bighorn sheep are fitted very well for life in their homelands high in the mountains. They are very sure-footed, because each foot, which has a cloven hoof, has a suction cup on the bottom. When they climb or go down a rocky slope or a sheer cliff, the suction cups cling to the rocks and give them secure footing. And their strong muscles take them in great leaps—fifteen feet or more, either up or down the mountainside. Watching them leap along the face of a sheer cliff, an observer expects to see them fall any moment and break their necks. They do not fall, but travel instead where most other living things cannot go.

Their coloring helps them, too. It is grayish tan with white underneath and a white patch on the rump. The overall color blends with the background of mountain soil and rocks so well that the sheep are hard to see when they are only a few feet

Rocky mountain bighorn sheep, ram and ewe, near Banff, Alberta, Canada

away. Yet their fine eyes let them see everything that is going on along the mountainside.

They have a rough coat of coarse hair about two inches long. Under it the skin is covered by a layer of very fine fur.

In the fall, the bighorn rams fight each other for the favor of the ewes. They run hard at each other, and their horns come together with a resounding bang. The force of the collision is almost enough to knock them out. In the larger rams, the tips of the horns are often blunted and frayed from these battles.

A bighorn ewe has a single lamb, or perhaps a pair of twins. The lambs are born in the spring, into a world of wind, ice, and snow, and often they do not survive. May is the most favorable month for them, but even then the mortality rate of the lambs is very high.

The ewe has her young in whatever shelter she can find—usually no more than a hollow in the rocks or under a ledge. The lamb can soon move about on tottering legs, and almost the first thing he does is nurse. At first he is small enough to stand up for this, but later on he grows taller and so has to kneel when he nurses. At first he nurses often, at least once an hour.

He and his mother stay in this place of dubious shelter for several days, and she stands guard against enemies. If an eagle flies over, she stands over the lamb, protecting him with her body.

In about a week, the lamb can follow his mother around, and they join the other bighorns of the band—perhaps as many as sixty mothers and their lambs. (The rams at this time go off by themselves.) When the lambs are about a month old, they can graze on tender foliage, even though they continue to nurse for many months. They are very playful, and play games with each other and their mothers very much like domestic sheep.

Their mothers eat two meals a day, one early in the morning and another in the afternoon, after which they lie down and "chew their cud"—regurgitate what they have eaten and chew it thoroughly before swallowing it again.

And always, always, they are on guard. Two ewes or more will stand guard duty while the others are eating or chewing their cud or resting; but the others, too, are watchful. To warn of danger they give a loud snort, and if an intruder continues to approach, they move away to a higher place—and may move, and move again, to avoid him. It is indeed a rare hunter, or photographer, who can come within range of them. They are protected by law in nearly every state in which they live.

THE DESERT BIGHORNS

The bighorns have various subspecies scattered through the Southwest and the Far West. One of these is the desert, or Nelson bighorn, in the mountainous or rocky parts of Arizona,

Desert bighorn sheep, ram and ewe, at the Desert National Wildlife Range near Las Vegas, Nevada

New Mexico, and Nevada. They are about the same size as the Rocky Mountain bighorn and have many of the same habits. They are sure-footed, with the same suction pads on their feet, and they have the same fine eyesight. They are a little lighter in color.

But they are exposed to more danger than the Rocky Mountain sheep. They cross from one high or rocky area to another, and in the crossing they are more in danger from predators and from the guns of man. However, they are quick to locate danger, and so they seem to be holding their own in numbers.

A herd of about thirty Nelson bighorns have been released in Zion National Park in Utah, in the hope that they will establish themselves in this area where they once were common. Sure-footed and nimble, they are very much at home in the rugged canyons and cliffs that characterize the area.

Another kind of mountain sheep, the Dall sheep of the Northwest and Canada, white in color, is closely related but is not regarded as a kind of bighorn.

A California bighorn

SUBSPECIES THAT ARE THREATENED

While the bighorns as a group are not endangered, there are two subspecies that are so listed. They are the California bighorn and the peninsular bighorn. The Badlands bighorn, that once lived in North and South Dakota, has become extinct.

The California bighorn is about the same size as the Rocky Mountain sheep, but darker in color. Its horns are a little smaller and more slender.

It roams the mountains of eastern Oregon and the high Sierra Nevada of California. In summer it goes to the crest of these mountains and along the ridges running west, but in winter drops into the lower eastern mountainsides. Northward, there are herds in southern British Columbia in Canada, and some of these may migrate into the high mountains of Washington.

Formerly, they were much more numerous both in Canada

and the western mountains of the United States. Today there are fewer than 200 in California, about 250 in Oregon, and a few in Nevada. British Columbia may have as many as 1,200.

Their numbers have been reduced by reckless hunting, by disease, and by competition for food with domestic sheep. To protect them, California has passed laws against molesting them, and that state as well as the National Park Service and the United States Forest Service, has restricted the use by man of some of the areas in which the bighorns range. In the Sierra Nevadas, in Inyo National Forest, the herds are protected in 41,000 acres devoted to use by them.

Numerous transplants have been made from British Columbia to the Hart Mountain Antelope Refuge and Steens Mountain in Oregon, with successful results. More recent transplants were made to the Sheldon Antelope Range in Nevada and to Lava Beds National Monument in California. It is expected that transplants will continue into areas where this bighorn once lived but has disappeared. Their homeland is being improved by cutting down the competition from domestic sheep, elk, and deer.

Like the other bighorns, both the California and the peninsular sheep are well fitted for life in their homelands. They are sure-footed, blend into the color of their surroundings, and are blessed with fine eyesight.

The peninsular bighorn is a little smaller than the Rocky Mountain sheep, and is even paler in coloration than the desert bighorn. It lives in the Santa Rosa Mountains and other parts of extreme southern California, and south into northern Baja California, in Mexico. Never very numerous, with a small range, it is threatened today with loss of some of its homeland to the development of home sites for people. It is estimated by the California Department of Fish and Game that there are fewer than 1,000 of this species.

Drought, hunting by poachers in spite of laws of many years' standing both in Mexico and California, and about 90 percent loss of lambs are reported as reasons why this bighorn is threatened. Also the subspecies interbreeds with the desert bighorn in areas where ranges overlap.

WHAT DOES THE FUTURE HOLD?

It is not beyond likelihood that the fate of other subspecies of bighorn will follow the route of the California and the peninsular sheep. In fact, they all may go. But everyone who comes in contact with these magnificent animals is moved toward strict enforcement of the laws, and protection of their homes and food supply.

It is believed that these efforts, continued by thoroughly aroused personnel in various agencies such as the national parks and the United States Forest Service, will insure the continued life of the bighorn sheep.

Chapter 9

ALMOST LOST

We were sitting on our neighbor's porch, with everything dark around us. We spoke in low undertones and kept very quiet. Off in the distance a whippoorwill called, and a nearer one answered him. These were the only sounds, except for a gurgle of water now and then from the little creek that ran near the porch.

Our host spoke, in a very low tone. "If they're here, they're sure keeping out of sight."

He was talking about a pair of beavers that had built a dam across the creek and made a pond. It was fine to have the beavers and their pond, but there was a problem. The place they had chosen to put the dam flooded the garden that we had made together.

We had torn the dam out time after time, but each time the beavers had put it back, in the same place. We had carried away all the sticks and stones they had used, taking them farther up the creek and leaving them, but the beavers would not follow our suggestion of another place to build. They simply brought everything back to that same spot and rebuilt the dam yet another time.

Finally, we had had the beavers live-trapped and carried to a place about fifty miles away, where they were released in a state park, in a little meadow, on another creek. They would have food there, and they would do no harm by making a pond.

They had been gone about two weeks, and there had been no sign of them around their old building site. We were congratulating ourselves that we had solved the problem. But today at dusk my neighbor had caught a glimpse of something in the water—something that might be a beaver. So we were sitting on the porch, keeping watch—but hearing nothing.

"I was probably mistaken," our host said at last. "I must have seen a shadow in the water. Mr. and Mrs. Beaver are probably working right now in their new pond, making their dam watertight." He got up to walk across the porch.

The minute his feet moved on the floor, we heard a sudden loud Whack! It sounded like a board brought down flat on top of the water in the creek.

We all recognized the sound. A beaver had brought his wide, flat tail down hard on top of the water, as a warning signal to any other beavers in the neighborhood. There could be no doubt about it. Our beavers were back!

We had to laugh, although, too, we had to admit defeat. We were not going to be able to move the beavers, short of killing them, and that we could not do. We would have to move our garden.

A DANGEROUS HABIT

This habit of the beaver, to cling persistently to his home pond, almost caused the animals to become extinct. When white men first explored the North American continent, they found beavers in most of it. Almost everywhere, except in the Florida peninsula and in the treeless Arctic North, were the ponds that beavers made by damming up small streams. Beaver fur was valuable, because it was used in the manufacture of tall hats that were a favorite among the fashionable men in Europe, and in rich, warm coats for both men and women. So beaver trapping had an almost unlimited market.

And the trapping was easy. If a man came to a beaver dam, the beavers were not frightened away from it. They merely submerged and waited until he went away. So the trapper could clean out every beaver that lived in the pond. All he had to do was make a break in the dam and set traps close to the break. The beavers came to mend the break and were caught in the traps. So trappers moved in great numbers through the continent's wild interior, cleaning out the beavers as they went.

BEAVERS ARE FITTED FOR LIFE IN WATER

Except for man, beavers have no enemies that could threaten them with extinction. They are wonderfully fitted for their life in the water. When they need to leave the water to get food, they can do that, too. In the water, beavers can outswim such enemies as bears, cougars, and bobcats.

A beaver has two kinds of fur. The top coat is made up of long hairs that the beaver combs with his toes, spreading oil on the hairs until they shed water and keep him dry. The second coat is a layer of rich, downy fur; it stays dry under the hairy coat, and keeps the beaver warm. So, even in ice-cold water, the beaver is warm and dry.

He is a fine swimmer, with strong hind legs and feet that have tough webbing between the toes. He uses his feet as paddles when he swims. He uses, too, his wide, flat tail to propel himself through the water and to guide himself wherever he wants to go. And when he brings his tail down hard on top of the water, the loud report warns other beavers of danger.

A beaver can stay underwater for as much as fifteen minutes. Thin valves of skin in his nose and ears close to keep the water out, and his eyes are protected with transparent skinlike lenses. His body functions, like circulation and digestion, slow down so that they take less oxygen. When he finally needs to come up for air, he can do it by lifting just the tip of his nose above water, so that he remains almost entirely hidden.

Beavers do not move as swiftly on land as they do in the water, but they can run fairly fast on their four legs. The front feet do not have webs and are like little hands that help them climb around on the dams that they build. They carry materials for the dam and pack them into the dam with their front feet.

Beavers are rodents, and so they have four strong, sharp front teeth. With them, a beaver can cut down a tree of considerable size, whittling it away a chip at a time. He sits up as he cuts, his front feet holding onto the tree and his strong tail bracing him firmly. And, of course, since he is a mammal and has lungs, breathing is no problem for him when he is on land.

A NEW HOME

When he is about two years old, a beaver usually leaves the pond where he was born and sets out to find a new place to live. He travels upstream or downstream or overland. Along the way he usually meets a female beaver that is also looking for a new home, and the two become mates, a tie that lasts through their lives.

The beavers travel along a stream until they come to a place where there are many small trees growing—willows and alders and quaking aspens. Probably they grow close to the stream's bank, and spread back from it over land that is fairly flat, but that rises a little as it spreads away from the stream.

The first thing the beavers do is decide where they want to build their dam. (Sometimes they do not build a dam, but hollow out a dry room high in the stream's bank. But most often they choose to build a dam.) At the spot they select, they cut a tree—and often more than one—and let them fall across the stream. Then they go upstream and cut more trees and brush, cutting the trees into short lengths. They float all of this down to the fallen trees and wedge it in to make the dam.

They dig out mud and stones, using their front feet to dig, and to carry these materials, hugged close to them, to the dam. They plug the mud and stones into the dam's foundation, and soon they have a watertight barrier that will hold the stream and make it spread out back of the dam to form a pond.

As the pond fills, the beavers go some distance back of the dam and make a great pile of sticks and stones and mud. This will be their lodge, where they can sleep and live in the winter-time when the pond is frozen, and be safe from enemies. The pond keeps on filling and water comes up high around the pile, but the top of the pile is still above the water. Then the beavers go underwater and dig a tunnel into the pile—a tunnel that starts at the pile's base and runs upward through it.

When the tunnel reaches the part of the pile that is above water, the beavers hollow out a room. They make a landing to go from the tunnel—which is full of water—into the room, which is high and dry for it rises above the water. They make the room big enough so that several beavers can be in it at one time. So here is a dry, safe home for the beavers, its walls so thick and sturdy that an enemy cannot easily tear them apart. It can reach the inside of the room only through the tunnel that starts deep underwater.

A FOOD SUPPLY

Beavers eat the buds and tender roots of water lilies and other plants that grow in water around the edges of the pond. But their main food is the inner, green bark of the willows and quaking aspen and other trees that grow near the pond. In summer, it is always easy to cut one of these down and eat the

Beavers building a lodge in Havasu Lake National Wildlife Refuge near Parker, Arizona. Nine beavers were observed at this lodge.

inner bark from it. For winter, the beavers can pile up a supply of food. They can do this because of the deep water in the pond they have made.

They cut tree after tree, and cut the trunks and branches into short lengths. They float these to a place in the middle of the pond not very far from their lodge. They make a big pile of them, anchoring the pile on the bottom of the pond. Then, if a beaver is hungry on a winter day, he has only to swim to the pile and pull loose a bark-covered stick. He takes it back to the lodge, and there he can sit, high and dry, and eat as much as he likes.

Young beavers are born in the spring, usually either two or four "kits" in a litter. They are very independent—covered with soft fur, open-eyed, and able to swim almost immediately. They are very playful, and have a great time teasing their parents and wrestling with older brothers and sisters. These older "babies" are probably born the year before, and stay with the family until the following spring. Then they leave the lodge and find a new place for a home.

A WHOLE COMMUNITY OF
PLANTS AND ANIMALS

Not only do beavers make a home for themselves that satisfies their requirements for land and water surroundings, but they provide the same kind of home for dozens of other kinds of animals—animals that would not live there if the beavers did not come first and build the pond.

Muskrats live in much the same kind of lodge that the beavers build. Frogs live entirely in the water at one stage in their lives, and lift their heads above the water to breathe air when they are adults. Dragonfly larvae live in the water and fly above it as adults; so do mosquitoes, on which the dragonflies feed. Ducks, kingfishers, and blackbirds are some of the birds that make their nests among the water plants that grow in the pond and feed on insects, plants, and fish that they find there. Some of this animal life might come to the stream even if it were not dammed, but with the pond there, the whole community of plants and animals is changed.

The beavers stay there, keeping their pond in repair, as long

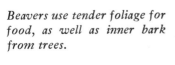

Beavers use tender foliage for food, as well as inner bark from trees.

as there is food for them. They even dig canals to trees that are quite a distance from the pond so that they can cut the trees and float them to it, rather than move. But the time comes when there are no longer any trees that will provide food. Then the beavers leave the pond and find a new place for their home. Gradually, the dam goes to pieces, and water drains from the pond. Slowly, it becomes a meadow with a stream flowing through it. Slowly, new trees grow. So, after a long time, this may become again a place where beaver can live, and a pair may select it for the place to start their home.

THE BEAVER'S COMEBACK

From the years when so many beavers were trapped, a few remained. Finally laws were passed to protect the survivors, and these reproduced rapidly. So today we have beaver dams widespread in many areas of North America (but still not in Florida or in the Arctic).

But there will never again be the great wealth of numbers that existed before the white man came to North America. He has built his cities and made his farms where the beavers once lived. The beaver needs small growing trees and a stream where he can build his dam.

Without these surroundings he cannot build a pond to live in —a pond that protects him from enemies and where he can store his winter supply of food.

Today a great many locations like this have been covered with the concrete and houses of cities or have been cleared of trees and drained, so that man can put them to his own use. So here, again, a wild animal has lost its great numbers because man has taken over the kind of place where it makes its home.

Chapter 10

AT HOME IN WATER

We lay very still on the grassy bank of a stream, watching a little pool just below us. The water was crystal clear, and on the bottom of the pool we could see a big trout resting, his nose upstream. He seemed to sway a little with the motion of the water, but otherwise he did not move.

Almost directly above him, a willow fly fluttered above the water. It came nearer the water, and suddenly the trout did move. He came swiftly upward, as if released by a spring. He arched out of the water, caught the willow fly, and dropped back into the pool, down to the bottom again. What a beauty he was!

We stayed still on the bank for a long time, watching him. Then, finally, we got to our feet. At our first movement, the trout was gone. With instant speed, and hardly seeming to move his body, he shot upstream, much faster than we could run.

The trout relied on his speed in the water to give him safety. Moreover, he could dodge quickly in the water; and he could hide in it. His body was colored in such a way that it blended with the shadows in the water and with the weeds and gravel on the stream bed, making it hard for enemies to see him.

And this place where he lived gave him food. Some of his food lay under the surface of the water—water insects and other small animals. Some of it, like the willow fly, came close

133

to the water from above and was snatched in midair by the quick-leaping hunter.

The trout's eggs are laid in water, and from the time they hatch, the fish spend all of their lives in water. They breathe with gills and so can take oxygen from the water. The water is a good home for them, and they do not come out of it.

THE PLIGHT OF THE CUTTHROAT

If the trout we saw had boasted a bright red splash across his throat, we would have been watching a very rare fish—one or another of the various kinds of cutthroat trout. In the early days of America, these fish lived in probably all of the high streams and lakes of the western mountains. Today they live in only a few of their former homes and are very rarely seen.

Reasons for their scarcity are several, important among them being pollution of their waters from mining and timbering. But probably the chief factor is the importation of other trout to keep the streams and lakes stocked for fishermen. These have not only competed disastrously for food with the cutthroats; they have interbred with them, and the pure strain remains

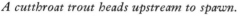

A cutthroat trout heads upstream to spawn.

only where other trout have not been brought into the area.

The Lahontan cutthroat trout is the biggest of the cutthroats, growing to ten to fifteen pounds. It lives in the Lahontan basin in California and Nevada, where there are pure populations in Independence Lake and Mackling Creek, California, and Summit Lake, Nevada, and a few tributaries. It is estimated that there are from 2,000 to 3,000. Lakes and streams where there are known to be pure populations are protected from stocking with others; and the Lahontan National Fish Hatchery of the Fish and Wildlife Service is devoted to the production of this beautiful fish. But it remains rare.

Even more rare is the Paiute cutthroat, with about 500 adults in the high Sierras of California, in Alpine County, considered "greatly depleted and in peril," by the U.S. Fish and Wildlife Service. Too, there are the greenback cutthroat, of which only ten pure specimens were known several years ago in high mountain streams of eastern Colorado; the Rio Grande cutthroat, of which a few live in tributaries of the Rio Grande in southern Colorado and northern New Mexico; and the Humboldt cutthroat, hardier than the others in regard to survival of pure populations among other strains, which lives in about twenty to thirty small headwater tributaries of the Humboldt River.

There are several others, in Colorado, in Montana, in Utah, and in Oregon, that are on the "status undetermined" list— species that may be threatened with extinction, but about which so little is known that their status cannot be determined.

SOME OTHER TROUT ARE ENDANGERED

The cutthroats are not the only kind of trout that are in trouble. In Sequoia National Forest in California, in Little Kern River and its headwater tributaries, the Little Kern golden

Lake trout damaged by the sea lamprey

trout lives—but only a few thousand in number, reduced by hybridization with rainbow trout, with which its streams were stocked. The Gila trout lives in Diamond Creek in the Gila National Forest, New Mexico—only 4,300 in number it is estimated, because its home was changed by foresting and by introduction of other species. The Arizona, or Apache, trout has been reduced from widespread distribution in the White and Black rivers to only Ord Creek and East Fork of White River and two small lakes in Arizona.

The Sunapee trout is found today only in Flood's Pond in Hancock County, Maine, where once it was widely distributed in lakes in New Hampshire, Vermont, and Maine. Its decline is blamed primarily on hybridization with introduced species. The blueback trout is another Maine species, living in eight lakes in the headwaters of the St. John and Penobscot rivers in northwestern Maine. It has diminished because of heavy fishing and hybridization with introduced species.

In all of these fish, deterioration of their home waters is one of their problems, as is competition from introduced species and hybridization with them. Steps to protect them include, wherever possible, cleaning up pollution as well as controlling the stocking of waters that may contain pure populations of endangered species.

TWO LEFT OF THE STURGEONS

Of all the unique fish we have, the sturgeons may be the oddest. A few years ago there were several kinds known to be in existence. Today there are only two. The shortnose sturgeon is a small species—somewhere around three feet long when mature—once common in all the rivers along the Atlantic Coast, but today seen only in the Hudson River. Pollution and overfishing are the factors that have reduced it.

The second sturgeon that we have today is the lake sturgeon, largest fish in the Great Lakes. It may grow to seven feet long and more, and to more than 300 pounds.

This sturgeon weighed more than 31 pounds.

The sturgeons were abundant over 100 million years ago, and have survived in much of their original form; so they are often called "living fossils." Their bones are more cartilage than bone, and they have heavy plates along their back and sides. They feed with the help of whiskers in front of their mouth, that drag along a muddy bottom and locate food. Then the mouth is thrust forward to engulf the food—small fish and other water animals that live in the muddy bottom.

There is an annual catch of lake sturgeon in the Great Lakes, varying from 1,000 to 3,000 pounds. This contrasts with millions of pounds taken in earlier years to secure the eggs for caviar, of which they were the principal source. They were also killed for the flesh, often smoked; and, because they destroyed fishermen's nets, were pulled out and destroyed as a nuisance.

Sturgeons go to the shallow waters of lakes and streams to spawn. The number of eggs varies greatly, depending upon the size of the fish. They have been counted from about 50,000 for fish of 11 pounds up to well over half a million for 112-pound fish. It takes the sturgeon about twenty years to reach maturity. It does not die after spawning as the salmon does, but returns to spawn again and continues to grow. A female does not spawn every year.

Sturgeons are protected today in most states in which they are found, particularly the lake sturgeon. Size limits allow it the long span of time that it takes to reach maturity, and because of this protection, the number seems to be increasing. In one sanctuary at least—Voyageurs' National Park—the lake sturgeon is completely protected.

CISCOES IN THE GREAT LAKES

In the first half of this century, the ciscoes were still a major part of the "chub" fisheries of the Great Lakes, spreading in millions of pounds through Lakes Michigan and Huron. There were the longjaw cisco, the deepwater cisco, and the blackfin cisco. The blackfin was the largest, the deepwater next; fifteen inches was considered a large size for any of them.

Like the cutthroats, they belong to the salmon family, and were seriously overfished for their delicious flesh. They were also attacked by the sea lamprey. Today no populations of these fine fish are known, and they are generally believed to be extinct.

BLUE PIKE HAS ALMOST DISAPPEARED

Another important commercial fish of the Great Lakes, especially of Lake Erie, in the first half of this century was the blue pike; the annual catch in Lake Erie was more than 20 million pounds. Deterioration of the habitat—the physical and chemical condition of the water and the biological habitat or relationship to other living things—is thought to be responsible for the nonsurvival of eggs and young. One example of habitat deficiency is a shortage of oxygen in the spawning area, after spawning takes place.

Very few blue pike have been seen in recent years. About 9,000 blue pike fry (very young fish) have been placed in the Gavins Point National Fish Hatchery at Yankton, South Dakota; fingerlings from there were stocked in an isolated lake in northern Minnesota. From these beginnings it is hoped that the blue pike species may be saved.

FISH ARE DIFFICULT

It is difficult to maintain an accurate check on the numbers of fish that exist, because it is hard to count them in water, and hard even to identify them. But the list of threatened wildlife of the United States holds the names of fifty-five threatened fish and twenty-three species of which the status has not been determined because of lack of information about them.

Many of these seventy-eight are species that have a very small population in some river or smaller stream or some lake, usually in the West; many have always had small numbers and sparse distribution. There are arctic graylings, for example, in remote streams and lakes in Montana, Utah, Wyoming, Washington, Colorado, and in Glacier and Yellowstone national parks; they are common in Alaska. Formerly there were three forms, the arctic, the Montana, and the Michigan; the Michigan form is now extinct.

It is thought there are a few of the Colorado squawfish, a big minnow that grows to a length of five feet, in the upper Colorado River and middle and lower Green River. This interesting fish was once widely distributed in the Colorado River and its major tributaries. It was decimated by the dams and reservoirs which changed its home from turbid, swift streams to a series of big lakes in which it does not reproduce.

Scattered widely over the country are many small populations of various kinds of dace, chub, suckers, minnows, pupfish, gambusia, darters, and others, about which little is known. Some of those listed are so seldom seen that they are thought to be extinct. Studies are continuing to locate the fish and to preserve some of their natural home, to protect them from extinction if they still exist.

In addition to those actually threatened or already extinct, many fish are being reduced substantially in number. No longer, for example, are flounder and haddock common in large numbers along the Maine coast, and no longer are there the great schools of herring that once went into sardine cans. Many towns that were once fishery towns have had to turn to other occupations.

On the other hand, appearance of Atlantic salmon in the Connecticut River is evidence that the program to restore this fine fish to rivers of New England is paying off. The work is being done by several of the states and the U.S. Fish and Wildlife Service.

Similar programs, for the rearing and transplanting of diminishing fish species, are going ahead in many coastal states and in big lakes, to preserve species of fish, many of which have been commercial headliner and prime sources of food in the past. These programs are supported by the states involved, and by United States grants to the states and to organizations within the states, such as a university or a combination of universities and other schools. The Mississippi-Alabama Consortium, for example, receives money from a national grant for research, education, and advisory services—funds for twenty-one projects and for thirty-five investigators from ten universities in the two states. Similar support is being given all over the country wherever the homes of fish exist in sizeable quantity.

Chapter 11

WHALES LIVE
IN WATER

If someone asked you to tell about animals that live in water, your first thought would probably be fish. But many other animals, too, live their whole lives in water, finding food and safety there and producing their young. Some are like the fish, and some are very different.

Many of these are animals without backbones—the invertebrates. They include some kinds of insects, some kinds of snails and their relatives, lobsters and crayfish, oysters, mussels and clams, and various others.

Among the animals with backbones—the vertebrates—there are many kinds besides fish that use water for a home, at least part of the time. Strangely enough, some mammals are among these—animals that have lungs and so must come to the surface of the water to breathe. Among them, too, are some reptiles—turtles, alligators, some kinds of snakes—as well as frogs, toads, and salamanders (amphibians). And fish have backbones and live all their lives in water, breathing underwater with gills instead of lungs.

WATER MAMMALS

Mammals that live in water include whales, porpoises and dolphins, and manatees; among them are the largest animals that we know. The blue whale, largest of them all, grows to be 100 feet long and to weigh 150 tons. These mammals are shaped much like fish, and swim like fish. The tail fluke (tail fin) is horizontal to the animal's body instead of vertical, as in fish.

These are animals that spend all their lives in water. On land for any length of time, they are likely to die because their skin sunburns or dries out. Because they breathe with lungs, they must come to the top of the water to breathe. We say that whales and porpoises "blow," shooting moist air through an opening just back of the animal's head, or on the back of its head. The moisture in the air condenses, so that the blowing is often accompanied by a fountain of mist. The blow not only can be seen, it can be heard. The different kinds of whales and porpoises make different kinds of noises, and so many of them can be identified by the sound of the blow.

There are various mammals that leave the water, some of them quite frequently. These include the seals and sea lions, the otters, the beaver. They breathe with lungs, but some of them can stay underwater for quite long periods. In some—the beaver, for example—the young are born out of the water. All of them look to the water for safety, able to move through it with great speed and agility.

WHALES, DOLPHINS, AND PORPOISES

These are the members of a large group of mammals—an order. They are divided into two suborders by scientists. If asked what these suborders are, the average person probably would say whales and porpoises. But he would be wrong. The toothed whales comprise one of the suborders, and they include many kinds of whales, of which we probably hear more about the sperm whale than any other. They also include the pilot whale or blackfish; the killer whale ("sea wolf"); and all the dolphins and porpoises. There are about seventy-five species of them.

The second suborder of whales consists of the baleen whales, which, instead of teeth, have baleen across the front of the mouth. Baleen is whalebone, and strips of it are attached to the roof of the whale's mouth and hang down like curtains across it. Across the lower portion, the curtains are divided into many threads, and form strainers. They strain the whale's food—fish and other kinds of small animal life—called krill, from the water so that the whale can swallow it without swallowing water. It is estimated that while their numbers were large, the baleen whales swallowed 150 million tons of krill a day.

The baleen whales are the "great whales," some of them the largest animals we have. The sperm whale, also a "great whale," has teeth. There are ten great whales.

All of the animals of the whale tribe live their entire lives in water. They mate in water, and their young are born in it, a year or more after the mating. Some of them travel long distances to calving grounds, where the young have the advantage of warmer water. The young are born one at a time, about every other year. Usually another female stands by at birth and

White whales grow to about 20 or 25 feet in length.

takes the baby to the surface of the water for its first breath of air. After that, it stays close to its mother, coming close beside or underneath her if it seems to be threatened.

Very little is known about whales that can be applied to all of them. They are streamlined, and the body is encased in a layer of fat, or blubber, just beneath the skin. In many, the body holds great quantities of whale oil that is very valuable commercially.

Apparently their hearing is their keenest sense, and they can hear echoes that they make by bouncing sounds off of objects or other animals in the water. They seem to have few enemies except for man.

THE KILLER WHALE

This is the only whale that displays any degree of savagery in securing its food. The others merely open their mouths and swallow whole any fish or other water animal small enough to be so eaten. But a pack of killers have been known to fall upon the largest whale and tear it to pieces.

There are many stories about them, some having to do with men in boats having disappeared in their vicinity. Some researchers discount such stories, but note that men of the sea—fishermen and sailors—are likely to feel uncomfortable when a school of killers is nearby. In addition to other whales, the killer feeds on seals, sea lions, and fish of all sizes.

The killer whale's body is black, and has a white splash running up the side that curves close to the tail, a white spot above and behind the eye, and white underparts. The animal grows up to about thirty feet long, with a large fin curving on its back, about midway between the head and tail. Killers usually travel in schools of five to forty. They range along both Atlantic and Pacific coasts, south as far as New Jersey in the East.

THE PILOT WHALE

The pilot whale, commonly called "blackfish," is as black as the killer and is about the same size, but is quite differently shaped. It has a blunt, high, rounded forehead that bulges forward, an adaptation that is thought to help in its search for the squid it uses for food. It serves as an efficient sound-echoing system at the depths that the squid inhabit.

Perhaps because it may become confused in shallow water, the pilot whale is the animal most involved when several become stranded on a beach. In recent years as many as fifty—an entire school—were stranded on a beach in North Carolina. It is thought that in such a stranding the whales lose their sense of direction and do not know which way to go to get back to deep water. So they follow each other in panic onto the beach.

BOTTLENOSE PORPOISES

Bottlenose dolphins, or bottlenose porpoises, are the animals we see most often in porpoise shows. They have most of the characteristics of the whale order: They're shaped like fish but have the horizontal tail fin; they navigate by echo guidance; they eat fish and other small marine animals; they come to the surface to get air through a blowhole at the back of their head.

Porpoises move in the ways of all the whales. They shoot through the water at high speed, with remarkable accuracy, and can turn swiftly, with no apparent loss of speed. They rise above the water in a smooth arc as they swim ahead or to the side, and they can leap straight upward through the water, lifting their whole length above it for several feet. They can dive at an angle or straight down.

Friendly and intelligent, a school of them in open water seems to like nothing better than traveling alongside a boat,

Mother and baby porpoise

crossing back and forth in front of it and showing off their smooth, arching glide. Their weapon against enemies, especially sharks, is the "bottle" nose. It is very hard and strong, and with it several porpoises drubbing a shark will make short shrift of it.

In captivity they are so intelligent that they have been taught countless tricks for exhibit, including affection for their trainers.

Their color is grayish with lighter underparts, and they grow to be about twelve feet long. They range the Pacific and Atlantic coasts, northward to California and to Cape Cod, and along the coasts of the Gulf of Mexico, as well as most of the other waters of the world. They travel in schools rather than singly.

There are a number of other dolphins and porpoises, similar in form and habit. The common dolphin, for example, is often

seen even in deep ocean water, playing around a ship and leaping high from the water. Its coloring is striking, with a bluish
back and side stripes of gold and white. The schools it travels in
are unbelievably large.

So many porpoises have been caught for exhibition that they
are not seen as often along American coasts as they formerly
were. Concern is growing that they may become a rare sight in
these areas. This is multiplied by the fact that they are caught
by the thousands in the nets of the tuna fisheries.

Scientists attempted to distinguish between porpoises and
dolphins, but the general public is so much inclined not to
make this separation that today the names are often used inter-
changeably. Most of the animals in "porpoise" shows that people see are, technically, dolphins—but nearly everyone who
sees them refers to them as porpoises.

THE GREAT WHALES

With the exception of the sperm whale, which has teeth, the
great whales are baleen whales. They range up to 100 feet in
length and weigh up to 150 tons. These are the whales of the
whaling industry. Because they have been hunted so ruthlessly,
eight of them are on the list of animals threatened with extinction. Together with the sperm, they are the blue, the finback,
the gray, the sei, the humpback, the right, and the bowhead.
The two that are not as yet on the endangered list are the
minke, the smallest, growing to 30 feet, and the Bryde's. These
two are forced to bear additional pressure because of protection
of the other species.

Whales are animals that roam the seas throughout the world,
and so their protection depends upon all the nations. But protection is hard to come by. Whale oil is immensely valuable, for
use as a high quality lubricant and in the manufacture of oil-

based products, such as many cosmetics. The oil, the meat, and other parts are used in food for pets and humans. One writer, David O. Hill in *Audubon* Magazine (January 1975) said, "Whales are very special creatures, and turning them into margarine, pet food, shoe polish, and lipstick displays unforgivable arrogance."

The International Whaling Commission was formed to regulate, among other things, the numbers of each species that can be taken each year. Eight of the big whaling nations of the world are members of the Commission and some of them have agreed to a ten-year total moratorium on all whale hunting, the United States leading the campaign for it. Others have supported movements in the Commission for curtailing quotas of various species. These include a recent cutback from 37,000 to 27,000 for the total whale catch, and a change in procedures that will allow the Commission to impose selective moratoriums for any species that falls below the level necessary for survival.

A spectacular example of the operation of the Commission shows a cutback for the finback whale, next in size to the blue (which is totally protected). A ban was placed on hunting the finback in the North Pacific and only 585 could be taken in 1975 in the Southern Hemisphere and North Atlantic—compared to the total for the preceding year of 1,550.

Prominent in the efforts to save the great whales is the Animal Welfare Institute of Washington, D.C. This organization's campaign includes the management of a citizen's boycott of Japanese and Russian goods, as these nations did not agree to the quotas of the International Whaling Commission. About 80 percent of today's commercial whaling is carried on by them. The boycott program was explained in newspaper advertising across the nation, and cooperating are some twenty conservation organizations and publications. Recent news from the Institute indicates that the boycott pressure is having a promising effect on attitudes of the two countries, and that the desired worldwide moratorium on the killing of whales may go into effect while there are still some whales of all species to protect.

The sei, the sperm, the minke, and the Bryde's are still hunted commercially. Cutbacks in the quotas for all of these were made by the Commission; two of them, the sperm and the sei, are on the endangered list.

All the great whales perform the acts of breaching and sounding. In breaching, the whale's whole body leaps from the water, falling back with a big splash. The humpback, gray, and right whales are best known for breaching. In sounding, which is very deep diving, the sperm whale takes the stage. It dives to very great depths to reach the squid that it eats.

THE SPERM WHALE

The sperm whale has teeth, and it is the only toothed whale that approaches the size of the baleen whales. The males grow to sixty feet in length and may weigh more than fifty tons. So the sperm whale is grouped with the baleen as a "great whale," even though it belongs to a different suborder. The average size of the males seems to be getting smaller, probably because the largest of them have been closely hunted. The young may be up to fourteen feet at time of birth.

The sperm whale is bluish gray with paler underparts. It has no fin on top, and its head is blunt and enlarged, with a thin, narrow lower jaw. The bulbous head contains spermaceti oil— sometimes a ton of it—valued as a lubricant. Hunting it for this oil, and formerly for ambergris once used in the manufacture of perfume, has reduced its numbers to about 25 percent of what they formerly were.

A very satisfactory substitute for sperm whale oil has recently been discovered in the United States. It is oil from the seeds of a hardy desert shrub, the jojoba. This oil serves in the same way as sperm oil, and is just as effective. It is hoped that with the establishment of plantations of this shrub and their maturity to produce seeds, that the pressure on the sperm whales will be lifted and that they will no longer be hunted.

The sperm whale dives deep, and can stay underwater for an hour and a half, or more. It can come up swiftly, without any sign of "the bends," a reaction that people get when changing from the varied pressures of deep water and shallow water. The sperm whale dives to get its favorite food, squid, although it also eats octopus, sharks and other fish, and other water animals.

The sperm whale is seen in greatest numbers where cold currents of water empty into warmer waters; such areas are often full of its favorite squid. The sperm whale lives in all the oceans of the world.

THE BLUE WHALE

Not only the largest of the whales, the blue whale is the largest animal man has ever known. It may exceed a length of 100 feet and a weight of 150 tons. It has been almost impossible to weigh one of these monsters accurately, and estimates vary greatly.

It lives in all the world's ocean waters, including the Atlantic and Pacific coasts, in widely scattered groups. The groups migrate to polar seas in summer and return to warmer waters as winter closes down.

The blue whale is slate or bluish gray with pale underparts, somewhat tinged with yellow. It has a tapered snout; its baleen is black, measuring up to forty-eight inches.

In the water it is spectacular, its spout throwing a fountain of water almost vertically and reaching thirty feet in height. When it dives, it often raises the tail flukes high in the air.

This whale is severely threatened with extinction. It now has complete international protection, but this was given only after its numbers had been reduced to only one percent of what they once were. The blue whale has not seemed to gain much after protection was given, still one of the scarcest of the great whales.

THE GRAY WHALE

Here is a fine example of a whale that came back from the verge of extinction, under total protection. The California variety is believed to have reached about 4,000 and is stable in number. However, the Atlantic variety has long been extinct, and more recently the western Pacific (Korean) strain is believed by some researchers to be extinct.

The gray whale is a good example also of migration carried on by these huge animals. Great numbers of them, traveling in rather small groups, move from Alaskan waters, which they leave in early winter. The whales travel down the Pacific Coast, and attract a big audience each year off the coast of California, as they move to the coastal waters of Baja California. Here are their calving grounds, and the calves are born over a span of a few weeks. They measure up to sixteen feet long and grow quickly, remaining under the protection of their mothers during the migration after their birth, in which they return to the Arctic. A female produces not more than one calf every other year.

The adult whales grow to a length of forty-five feet and weigh about thirty-five tons. They are grayish black and are mottled with splotches of barnacles, and have yellowish baleen, twelve to fifteen inches long. Their blow is not as high as in some other whales—only about ten feet; the flukes often come up over the water as they dive.

THE FINBACK WHALE

The finback is next to the blue whale in size, growing to seventy feet long or more and about seventy tons. The young is about twenty-two feet long at birth.

This whale is gray above and white beneath, with many grooves or folds beneath, from the tip of its mouth to more than halfway to its tail. It has a small fin on the back, near the tail. The bluish gray baleen is colored with irregular streaks of purple and white, falling to a length of thirty-six inches.

Spouting, it throws a narrow column of water fifteen to twenty feet high, and sends out a loud whistle. It is found in all oceans, including coastal waters of both Atlantic and Pacific off North America.

THE SEI WHALE

A close relative of the finback, the sei whale (sometimes called rorqual) is very similar, except smaller and darker. The fin on its back is larger and nearer the middle of the back. The baleen is black or gray, and is about two feet long. It blows a column of water up to eight feet high. The sei may grow to fifty feet or more. It occurs in all the world's oceans, and is seen off both the American Atlantic and Pacific coasts.

THE HUMPBACK WHALE

Growing to forty to fifty feet, the humpback is like the finback in having a tapered snout and a fin on the back; it is a little nearer the tail than midway, and there is a smallish bulge between it and the head. It has long side fins starting just back of the mouth and continuing more than halfway along the body, slanting downward.

The humpback is black with white throat and underparts. The baleen is black. The spout is a high one, reaching twenty feet. The humpback occurs in all oceans.

THE RIGHT WHALES

There are two species of right whales, the Atlantic and the Pacific. The Pacific is the larger, growing to seventy feet, compared to about fifty-five in the Atlantic. Both are blackish, with a bulbous head that has black baleen about eight feet long. The spout makes two columns in the form of a V, rising to ten to fifteen feet high.

The right whales occur in polar, temperate, and tropical seas, in the western Atlantic Ocean and the eastern Pacific.

THE BOWHEAD WHALE

Here is a whale that clings to Arctic ice, occurring only in polar and subpolar seas. It is shaped a little like the right whales, but its rounded head is nearly half its length. It grows to sixty-five feet. Its spout, again like the right whales, rises in two columns, ten to thirteen feet high. The baleen is black and grows to more than ten feet.

The bowhead can crash through ice floes to blow, and can stay underwater an hour or more, which fits it for life in the ice-filled water that is its home.

WHAT ARE THE PROSPECTS FOR THE WHALES?

No one seems to hold out much hope for survival of all the species of great whales. The wheel may be very slowly turning in their favor, but the question is whether it will reach the degree needed for survival in time for one or more of them.

Millions of great whales have been killed—millions, in fact, in this century, as one after another of the great herds have been decimated. All of them are on the endangered list except the minke and the Bryde's. Yet many have received some measure of protection for many years and have made no comeback. Chief among these are the bowhead, the right, and the blue.

On the other hand, the California gray whale, given protection, has come back encouragingly, building to a safe and stable population. What it has done, others may eventually do.

It is, however, a race between survival of the whales and survival of the whaling industry. Reading the inexorable end of the industry in the falling source of supply, some whalers are inclined to grab what they can while it is there, and before their whaling equipment has to be replaced at exorbitant cost. So it becomes a matter of which will go first, the whales or the equipment.

In the meantime, more and more protection is being provided through the International Whaling Commission. The recent protection given the finback is a major victory in this direction. So it does not seem too much to hope that at least some of the great whales will continue to roam our oceans.

THE FLORIDA MANATEE

The manatee is another animal that lives entirely in water. Like the whales, it is a mammal and must come to the top of

The clumsy-looking manatee lives entirely in the water, bearing its young there and feeding on water plants.

the water to breathe, thrusting its nose above the surface and taking in a big gulp of air. But it is not a whale and is not related to them, falling into an entirely different order.

It once ranged in fairly large numbers in warm coastal waters from North Carolina around the coast of Florida and along the Gulf Coast to southern Texas. It was shot for sport, and for the meat and its skin and oil. The number was further reduced by sudden cold spells in the weather; when the water chills, the manatee is very susceptible to pneumonia. Also it is frequently run into and fatally injured by power boats.

Today its range is greatly reduced, but it is still found in small groups at various points along its former range. It can be seen at Everglades National Park, which has the largest sanctuary protecting this animal. It is also seen at several national wildlife refuges, including Chassahowitzka at Cedar Key, the

J. N. "Ding" Darling Refuge at Sanibel Island, and Merritt Island Refuge at Pelican Island, all in Florida. It is not known how many exist, but the animal is much less numerous than in earlier years.

A single manatee calf is born at a time, in the water; it is not known how often the cow gives birth. The manatee is a long, awkward oval weighing up to 1,000 pounds. Its forelimbs are a pair of flippers; it does not have hind limbs. The tail is flattish and rounded like that of a beaver.

The manatee feeds on vegetation that grows in the water, sometimes entirely freeing the water of such plants as the water hyacinth. It is a sluggish animal, eating its way slowly through weed-choked water, and could be exterminated quickly if it were not protected. It is protected throughout Florida.

Close cousins of the Florida manatee live in the waters of the West Indies and along the coasts of Central America and northern South America.

These are the mammals that live strictly in the water. They find safety in it; they find food in it; they produce their young without coming out of the water. They are animals to which water is truly a home—one that they never leave and that they must always rely on to produce their life necessities.

Chapter 12

IN AND OUT OF WATER

The egg from which he hatches is laid in the water; it is one of a cluster of eggs, each surrounded by a globule of clear jelly. He can be seen before he leaves the jelly—a tiny creature with a distinct fishlike body shape, a blunt nose, a tapering tail. He swims free of the jelly and swims in the water, taking in food from it, never coming out on land. He is a tadpole—a water creature. He does not even come above the water to breathe. He can take oxygen from the water through tiny gills.

Time goes by, and he gets two front legs, and two hind legs. His tail disappears. He climbs out of the water, onto water weeds or the bank of the pool or pond where he was a tadpole. Now he breathes with lungs, not gills. He is a frog.

AMPHIBIANS

All of us are familiar with this kind of change from a water home to a land home. The frog does not entirely give up water as a home. We have watched him dive back into the water from land, seeking safety. He can swim in it, and stay submerged in it for quite a long time. But he cannot change again to an animal that lives entirely in water. He breathes the open air, and gets some of his food from it.

Animals that make this kind of change are amphibians. They

have backbones—frogs and toads, salamanders, efts, and newts. There are amphibious insects, too—mosquitoes and dragonflies, for example. Their eggs are laid in the water, and after the eggs hatch, they grow through the first stage of development as water creatures. Then the body of the insect changes until it is fitted to live in the open air; it can take oxygen from the air, and it can fly in the air or crawl on land.

MAMMALS

This is not the only way in which animal life moves from water to land. There are many animals, some of them mammals, some reptiles, that leave the water more or less briefly for various reasons, and return to it again, without a change in the structure of their bodies.

Otters are mammals that use both land and water for their home. River otters are found in the rivers and ponds of most of North America north of Mexico; sea otters live at the edge of the ocean along the Pacific. Other water-land mammals include seals, sea lions, and walrus.

SEA OTTER

Of all the mammals that leave the water and go back to it, sea otters probably leave it the most briefly. Some of them, perhaps, do not actually leave it at all. This animal is almost never seen on shore. It prefers the shallow water in and around a bed of kelp, where it spends most of its time floating on its back. Where a walrus will pull out on the beach and lie sunning itself, the sea otter wraps a length of kelp around itself so that it will not float away, and lies on its back on the water or in the kelp, perhaps asleep.

It is an animal perhaps four feet long, including the tail, and weighs up to eighty-five pounds. Its rich, dark brown fur is beautifully highlighted with white and gray, heavily accented in the head, which has whiskers and a comical effect of an old man. The front feet—"hands"—have toes that are almost as agile as fingers, while the large, strong hind feet are heavily clawed and webbed. It is a fine swimmer, diving and dodging, darting this way and that without effort, and can hold its breath underwater. It does not find it too difficult to hide in the kelp from its natural enemy, the killer whale.

It feeds in the water, diving to the bottom to come up with an abalone or sea urchin or other water animal. It floats on its back and holds its prey on its stomach, picking it to pieces with its "fingers" and eating it in little bits. If it has a hard shell like a clam, the otter may bring up a flat stone and, the stone balanced on its "stomach-table," pound it to pieces against the stone.

The young are born singly, probably in late spring. For their birth the mother may emerge onto the rocks of a small seashore inlet, or may go to a heavy, protected bed of kelp. She cares for her baby with the greatest of devotion, nursing it, guarding it, playing with it as it rides on the table that, upside down in the water, she makes for it. She will defend it to the end, often giving up her own life in the process.

There are two big colonies of sea otters. One is in the Aleutian Islands, Alaska, where there are thought to be several thousand animals. The other is off the coast of California, where more than a thousand have been estimated.

The sea otter was once numerous all up and down the Pacific Coast, but was slaughtered by the thousands for its rich fur. It now receives state, federal, and international protection, so that we can hope it will someday enjoy again large numbers along the coast.

Sea otters leave the water briefly to sun themselves on a small group of rocks.

RIVER OTTER

Although the river otter is closely related to the sea otter, it is quite a different animal. No floating around upside down on the water for this one! It moves through the water like a fish, thrusting its nose up to breathe. And when there is something beyond the water that it wants to investigate, it leaves it readily, traveling easily on all fours. When it wants to sun itself or just rest, it lies on the bank of a pond or stream.

It is six inches to a foot shorter than the sea otter, and more slender; it, too, is dark brown with lighter markings on underparts. Its snout is somewhat tapered but rounded at the tip, and the whole head is tapered rather than rounded as in the sea otter. Both front and back feet are webbed.

River otters are probably more at home in water than they

are on land, returning quickly to the water if they are threatened. They get their food in water—fish, frogs, crayfish, and other small animals that live in water. But they will leave one stream and travel overland to another if they need to find a new location.

The den is above water, dug into a bank or hollow and reached by a passage through the water. Here the young are born, blind at first and covered with fur; usually there is a litter of two. The young are very playful, a characteristic that continues throughout the animals' lives. Only when it is frightened and driven into hiding will the otter abandon the pastime of some game or other, playing with other otters or perhaps only with rocks or sticks.

River otters are found in almost every state in the United States, including Alaska, and throughout all but the northernmost parts of Canada. They easily make use of whatever waterland home an area offers, and so have established their kind far and wide.

SEALS ARE WATER MAMMALS

There are two great families of seals, the fur seals (this family includes sea lions) and the hair seals. Seals spend most of their lives in water but leave it frequently, coming out on the ice or on a beach. They breathe with lungs. Their legs are flippers, which are very much more efficient for moving in the water than on land. They feed in the water, catching fish, squid, and other small water animals. If threatened, they return to water for safety.

The fur seal is like the sea lion and the walrus in that the strong hind flipper can be turned forward and used as if the animal were walking on it. So when it leaves the water, it moves along with a lurching gait in which all four flippers are used.

In the hair seal, the hind flipper drags behind when the animal moves on land. Its motion is one of wriggling across the sand or ice, rather than an approximation of walking.

Several seals are on the list of endangered animals, their numbers severely reduced by man, who wanted their soft, lovely fur. Mother seals have only one baby in a year, or perhaps every two years, another reason why their number is likely to decrease. They come out on land or ice to have their young, a time when they are very vulnerable.

One of them, the Caribbean monk seal, is thought to be extinct; at best it exists in only small numbers. One very much like it is the Hawaiian monk seal, living in the vicinity of the Hawaiian Islands. It is protected by state and federal law. Most of these seals, which are estimated to number about 1,000, live within the Hawaiian Islands National Wildlife Refuge. The female monk seal measures about 7½ to 8 feet and weighs an estimated 600 pounds; the male is a little smaller.

Other seals that are threatened are the ribbon seal and the Guadalupe fur seal. Little is known of the ribbon seal except that it lives in the Alaska area, northward in the Bering Strait. It is a much smaller seal than the monk, the male weighing about 200 pounds and measuring about five feet long. It has light bands through its brown fur, around the neck, around the front flipper, and around the back part of the body, toward the hind flippers. No estimate is made of their number, although it is known that they have a very small population. There is no provision for their protection.

It is estimated that there are about 1,000 of the Guadalupe fur seal, living in the region of Guadalupe Island, Mexico. They are protected by the Mexican and California laws, and it is hoped that their gradually increasing population will continue increasing and will extend to the California coast.

The Alaska fur seal is the leader among seals that are commercially important, with a harvest each year among the older

The Alaska fur seal uses all four flippers to "walk" on land, lurching along with a clumsy gait that is replaced by swift grace when it enters the water.

A baby fur seal suns itself but does not go far from the water.

Stellar sea lions. A big bull guards his "harem" of several cows on rocks in the Aleutian Islands National Wildlife Refuge.

males. Among the hair seals, the harbor seal is perhaps seen the most often, along both the Pacific and Atlantic, at the mouths of rivers and in shallow harbors. It frequently comes onto land.

THE SEA LION

Belonging to the same family as the fur seals, the sea lions are very much like them. There are two kinds, the northern, or Stellar, sea lion, and the California sea lion. With a body much like a fur seal, the sea lion has four flippers, and the hind flipper can turn forward for "walking" on land. The northern kind is very much larger than the California. The northern male measures up to 10½ feet and weighs to 2,000 pounds, while a California male is large if it reaches 8 feet and weighs 600 pounds. The females are smaller, weighing up to 600 pounds and 200 pounds respectively.

The northern kind is a light brown, while the California is a medium or dark brown. Common enemies are the killer whale and large sharks.

Sea lion young are born a single pup to a female. Before the birth of their young, the females come onto shore or to an island. Soon they are followed by the males, who fight over "harems" of the females. The result is a collection of males, females, and young that keep up an almost constant milling about.

The young are a darker brown, and have big, blue eyes that later change to brown. They soon learn to swim, and are active in keeping up with their mothers. When, after the birthing season, the adults abandon the nursery shore, the young are quite able to go along with them.

The northern sea lion ranges along the Pacific Coast from northern California northward to the Aleutians and Bering Sea. The California variety ranges the Pacific Coast from Monterey Bay northward, sometimes as far north as British Columbia. It is the California variety that is often seen on Seal Rocks off the coast at San Francisco.

The California sea lion is often called "seal" at aquarium shows along the Pacific Coast. It can be trained readily, and its loud, almost continuous barking is well known to its many fans. The true seals do not keep up this continual barking, and the northern sea lion is usually quiet.

THE WALRUS

This big animal is like the fur seal and sea lion in the general shape of its body, but would never be mistaken for a seal. Both male and female have big tusks of fine ivory. They have "walk-ing" flippers, both front and rear, and waddle about on a beach much as do the sea lions. But they have no fur, and only a few hairs show this mammal characteristic.

Wrinkled front and shoulders are reared high as they move about on land. They disappear underwater if they are startled; they are good swimmers and can soon leave an area of on-shore danger. Sharks and killer whales are almost their only enemy in water.

Baby walruses are born out of the water, on ice or on shore. Each female has only a single pup, which she devotes herself to caring for. If one is attacked, other walruses may help in the defense; in fact, they may come to the assistance of each other if some are in danger.

They sometimes congregate in great numbers on ice, on an island, or on the land. They live in arctic waters, coming southward in the winter but never leaving the ice for long or for a great distance.

REPTILES IN WATER

Many reptiles seek the water and spend a great deal of their lives in it. Some of the world's biggest snakes are fine swimmers; they can move faster in water than on land, and so are safer there. They find much of their food in water.

Many small snakes, too, are at home in water. There is one called a "water snake," although he goes easily to land and is often seen there.

ALLIGATORS AND CROCODILES

Among the reptiles most closely associated with water are the alligators and the crocodiles. They are much alike; but the American crocodile is much rarer than the alligator. It is thought to be declining slowly in Florida and probably throughout its range, which includes southern Florida, some of the Caribbean islands, the Atlantic Coast from Yucatán to

Colombia, and the Pacific Coast from Sinaloa to Ecuador. It is on the endangered list.

The American alligator has become quite common in the waters of most of the Gulf states and the southern Atlantic states. Legal protection that makes it unlawful to ship alligators, their hides, or products made from hides, across state lines has erased the former heavy threat of extinction, and the alligators have come back so hardily that they are becoming a nuisance in many communities that are near swamps or lakes. They are frequently caught and moved to a location more remote from people.

Alligators and crocodiles have the same heavy, horny skin. They rely on water in much the same way, returning to it if startled, getting their food—fish and other water animals—almost entirely in water.

But they leave the water to make a nest in reeds above a swamp or on the bank of a lake or stream. Their eggs are laid in this dry haven, and the female reptile stands guard nearby. When the eggs hatch—anywhere from fifteen to eighty-five—the young reptiles make straight for the water. There the

American alligator

mother continues to stand guard, although many of the young are lost. With the crocodiles, many eggs fail to hatch.

Alligators and crocodiles breathe with lungs, so must come to the surface of the water for their oxygen supply.

TURTLES SEEK WATER

Turtles are another kind of reptile that is seen more commonly around water than on dry land, although many of them seem quite at home in both areas. The turtles in a swamp or lake often leave it for a half-submerged log, and spend their time sunning in the open air. They breathe with lungs and must come to the surface for a fresh supply of air, although they can stay underwater for long intervals.

They get their food from water—insects, fish, certain kinds of water plants. But they come ashore to lay their eggs. When startled, they plunge quickly into the water and dive deep to hide among the water plants.

BOG TURTLES

The bog turtle is a rather small turtle that lives in colonies in freshwater marshes and meadows from Connecticut to southwestern North Carolina. There are estimated to be only about 500 colonies, and the bog turtle is on the endangered list. It can be identified by a large yellow or orange patch on the side of the head, and has no yellow spots scattered on the shell.

The eggs are laid on land, probably not more than three to five eggs to a nest. It is against the law to keep one of these turtles as a pet. Extensive collections of them for sale in pet shops contributed to the decline of this species, although the primary cause was destruction of habitat for cultivation and for housing development.

Left: Baby turtles released near the water in "Operation Green Turtle"

Close-up of a baby green turtle

Below: A sea turtle deposits her eggs in a hole she has made in the sand. This is the only time she leaves the protection of the sea.

SEA TURTLES

There are several different kinds of sea turtles, all of them large, and all of them coming on land only to lay eggs. Largest is the green turtle, which grows up to 1,500 pounds. It lives in tropical oceans, and nests commonly in the Hawaiian Islands. Once it nested along the southern Atlantic seaboard of the United States, but was practically wiped out there.

The green turtle is said to be the most valuable reptile alive. It has been hunted by man for many years, for its meat, eggs, and calipee, the cartilage, used in soup. Its skin has been used for leather, its shell for jewelry, and its oil for cosmetics. But it is now protected by law in many of the areas where it once was taken, with laws against trade in various products.

Additional aid to the green turtle is given in the artificial hatching of its eggs and protected release of the young turtles. Ordinarily these are snatched up in great numbers as they move along the beach, heading immediately after hatching for the open ocean. But many die before they reach the water, caught by gulls and other birds.

Man has for a number of years carried on "Operation Green Turtle," in which the eggs are moved to a safe place and hatched in an incubator, then carried to water deep enough to protect the baby turtles. There they live, eating turtle grass that grows near the beaches, and migrating through the water sometimes for great distances, until they are adult turtles. They eat in the water and breed in the water. Then the females return to the beaches where they were hatched to lay their eggs.

Reports came in during the summer of 1975 of green turtles coming to Florida Atlantic beaches to lay eggs, and it is believed by researchers that they are some of the young turtles that were released there. Reports have been sent in that as many as six to eight a night have been seen on the sand. They are rigidly protected, and may be the nucleus of a new off-Florida population. This is a big change in their home in favor of the turtle, in that they are not disturbed there, and even their eggs may be safe from marauders.

Another kind of sea turtle, the loggerhead, is receiving protection from man. Sanibel Island, off the Gulf Coast of Florida, has long been a central nesting ground for these turtles, which grow up to 450 pounds. But much developing on the island has brought homes, hotels, and other buildings, and people with them, and the turtles have been in danger of being crowded out. So a group of more than 400 people work together during the nesting season, to protect the turtles and their eggs. Other groups work along the Gulf Coast and along the Atlantic Coast, to perform the same protective action. So it is hoped that Florida will not lose its loggerheads as it did the green turtles.

Other sea turtles, all with many of the same habits of life, are the Pacific ridley, the leatherback, the hawksbill, and the Atlantic ridley. They all contribute to the record that the sea turtles set for spending most of their lives in the water, although they breathe with lungs and must come to the top of the water to get air. But they—and the females only—actually come out of the water and move onto land only when they lay their eggs.

Chapter 13

HOMES ARE CHANGING

Suppose that part of a man's field is swampy land that holds too much water for him to use as a field. He does not think that many different kinds of animals are using it for a home.

A muskrat lives in the water, putting its head above water to breathe and coming out in a "house" that it has built in the water. The house is built of sticks that he has brought here, and it rises, a pile of sticks, above the water. It is hollow inside and has a platform above the water. The muskrat eats here, as well as eating plants that grow along the edges of the little bog. He brings his mate here, and here the young are born and are raised until they are large enough to leave the house and play in the watery surroundings of the house.

Dragonflies live in the water of this little swamp. From eggs in the water they hatch to larvae and later leave the water to fly in the air as full-grown adults. Mosquitoes do, too. Blackbirds live close to the water, building their nests in cattails that grow there. Crayfish live in the muddy bottom. Turtles crawl out on the edge and sun themselves.

For all these animals there comes a big change in their home. The man decides to drain the little bog, never knowing that he will destroy the homes of many animals. He lays tile in the bog, and water runs through the tile and into a ditch. The bog becomes dry, and now he can plow it with the rest of his field.

What happens to the animals? Some of them die. Some of

them move along to another swamp, if they are lucky. But even moving is not a simple thing. The swamp they move into may have all the animals it can take care of. So, gradually, some of the animals die, until again there are only enough for the swamp to support. There must be food for them if they are to live there.

And so it goes that, as animals' homes are destroyed, the number of them is reduced. If enough homes are destroyed so that the number of the animals becomes very small and is still decreasing, they show up on our list of threatened or endangered animals. They are threatened with extinction.

SAN JOAQUIN KIT FOX

This is a desert-dwelling fox of which it is estimated there are only between 1,000 and 3,000 alive. It weighs only about five pounds, and is a buffy yellow with a black-tipped tail. It once ranged most of the San Joaquin Valley, in California, but today is seen, only occasionally, in five counties on the west side of the valley. It moved to this west side from the rest of the valley

Kit fox

As man occupied the wilderness, he left less and less room for the grizzly bear, which is found today only in remote areas.

as it was cultivated and homes for people were built there. The areas that it moved into still are undisturbed in their natural shrubs and other vegetation.

It is illegal to hunt and trap this little fox, although some poaching may still be going on. It is proposed that attempts be made to reestablish the animal in areas that still offer a suitable home.

GRIZZLY BEAR

The United States Fish and Wildlife Service has recently placed the grizzly bear on its list of threatened species, following a survey that indicates there are only between 500 and 1,000 of the big bears living south of the Canadian border. This means that in the forty-eight states south of the border, the bear is protected from hunting and trapping—from harassment of any kind, unless there is an immediate threat to human life or serious inroads on livestock.

The grizzlies require larger areas of wilderness than any other North American mammal. The territory in which they live is divided into "ecosystems." These are the Bob Marshall Ecosystem, in Montana; the Yellowstone Ecosystem, which is made up of segments of Wyoming, Montana, and Idaho; and the Selway-Bitterroot Ecosystem, segments of Idaho and Montana. In Montana, twenty-five bears may be taken from the ecosystem each year.

The primary reason for the decreasing number of bears is thought to be the increasing occupation of wilderness area by man, who puts it to various uses and destroys its wilderness character. This causes the bears to die out, or to move north of the Canadian border. North of the border the grizzly population is still considered to be holding its own, as the snow and cold reduce the number of people who live there.

MEXICAN DUCK

The Mexican duck is another animal that is suffering from a loss of its homelands. It is native in southern New Mexico, Arizona, and Texas, wherever there are marshlands and small lakes. But the swamp north of the border is being drained, and the number of ducks has fallen to only about 500. South of the border there are more, reaching perhaps 1,500. Accompanying habitat destruction is a tendency to hybridize with mallard ducks, which are working southward to mingle with the Mexican species.

The Fish and Wildlife Service is working with the New Mexico Department of Game and Fish to restock areas where the duck formerly lived, and has released more than 100 of them in the Bosque Del Apache National Wildlife Refuge in New Mexico. It is hoped that this will furnish a center for a new and large population in this area.

MASKED BOBWHITE

Here, again, is a bird, the masked bobwhite, that formerly lived north of the United States-Mexico border and is now being released north of the border in an attempt to reestablish it there. It lived close to the border in Arizona, where tall grass grew intermingled with mesquite and cactus. The bobwhite depended upon the grass, and, slowly, the grass was eliminated by the grazing of cattle. So the bobwhite, too, was eliminated.

The masked bobwhite is quite similar in appearance to the bobwhite of the eastern United States. But it is smaller, and the male has a brick-red breast, black head and throat, and a white line over the eye. The female is very similar to the common bobwhite. There are only about 1,000 of these little birds.

Destruction of grass in their homeland caused the disappearance of the masked bobwhite.

Some of the birds are being raised at the Patuxent Wildlife Research Center of the Fish and Wildlife Service, in Maryland, and are being released in Arizona habitat like that in which they formerly lived. Meanwhile studies are being made of the bird's life history and its range both in Mexico and the United States.

MORE DESTRUCTION OF NATURAL HOMES

Various other birds (and other animals as well) have suffered because their homes have been destroyed. One that we hear a great deal about is the big ivory-billed woodpecker, which is thought by many researchers to be extinct. It was once widespread through the southeastern United States, but this big bird has not been seen recently. The most recent sightings have been in southeast Texas, southern Louisiana, and central South Carolina. The big white bill and large patches of white on the wings of this big bird of woodpecker habits unmistakably identify the ivory-bill. But its numbers have been reduced to very few and perhaps none at all because its home has been cut from under it. It formerly lived in over-mature forests where dead and dying trees held the larvae of the wood-boring beetle upon which it depended for food. Timbering out the old trees was a major project before the woodpecker's dependence on them was known.

It is strictly protected by law, and search for it is being continued, as well as for habitat that would make a suitable home.

Another bird that has followed the same pattern is the red-cockaded woodpecker. Much smaller than the ivory-bill, it has black and white horizontal stripes on the back, and a black cap and stripe on the side of the neck. The male has a small red spot on each side of the black cap.

This woodpecker once lived throughout southern Missouri,

western Kentucky, and southeastern Virginia, south to the Gulf Coast and southern Florida. But today, with only an estimated 3,000 to 10,000 surviving, it lives only in a few old pine woodlands that still stand in this area, and is very vulnerable because the bird must nest in this kind of woodland. It, like the ivory-bill, depends upon diseased trees—trees infected with redheart disease—to give it insects for food, and many of these trees have been cut and disposed of. Today's efforts to protect the woodpecker include measures taken to leave a few of these trees standing to serve as the woodpeckers' homes. A group of scientists, working as a "recovery team," is researching the problems that must be solved if the woodpecker is to continue to have a home.

Another bird that has lost many of its natural homesites is the burrowing owl, that lived, along with its owner, in prairie dog holes. It is rated as "status undetermined" because not much is known about it in the areas where it still can find burrows to live in. But it is known that many of these burrows were destroyed as the prairie dogs were eliminated, and so the burrowing owls went, too.

WOLVES AT THE BORDER

Two kinds of wolves live in various areas along the United States-Mexican border. Both are very rare and becoming more so, their numbers losing ground because of changes in their natural homes.

One is the red wolf, that once roamed the southeastern United States from Florida to central Texas and north as far as southern Indiana. Today it is gone from most of that range. It hybridizes easily with the coyote, and many of the animals that are seen are part coyote. There are a few small pure popula-

The red wolf

tions in four southeastern Gulf Coast counties of Texas and adjacent Louisiana.

The red wolf looks much like a large, coarse-furred coyote; it is hard to distinguish it from the coyote.

As people moved westward and built their homes in its territory, the red wolf found it hard to adjust. The coyote adapted more readily, and so pushed deeper and deeper into red wolf areas, crowding out the red wolf and hybridizing with it. So today red wolf numbers and range are greatly restricted. It is one of the threatened animals that a recovery team, made up of Louisiana, Texas, and Federal researchers, is working on to try to protect from being killed and from hybridizing with the coyote.

The second wolf sometimes seen along the Mexican border is

the Mexican wolf, smallest wolf in North America. The male weighs less than 100 pounds, the female still less. It is quite dark, with a short, broad face.

The Mexican wolf was thought to have been wiped out in the United States, where it lived in southern Arizona and west Texas. It was hunted unrelentingly as a danger to livestock, until it was not seen north of the border. But in recent years it reappeared along the border and is protected now by no-hunting regulations.

ALONG THE EDGES

The red wolf is a good example of quite a large number of animals that are seen rarely inside the United States but more often across the borders of Mexico and Canada. These animals are called peripheral, meaning that they live along the edges of this country but spread deeply into bordering countries.

The jaguar is a good example of an animal that is peripheral between Mexico and the United States. It is seen, although rarely, from central Texas south to Mexico, and more commonly in the Mexican part of its range. In roughly the same range, some of its cat cousins are also sighted—but, again, rarely north of the border. These are the jaguarundi, the ocelot, and the margay.

The coatimundi is another peripheral in this area, an animal that is a close cousin of the raccoon. Its color is a grizzled brown; it has a long snout and a long tail, the head and body averaging about two feet and the tail another two feet. The tail, often carried erect when the animal is on the ground, has indistinct rings. The head, marked with black and with white over the eyes, looks a little like a raccoon. It is larger than the ringtail, or ringtail cat, which lives in much the same peripheral

The ringtail is a peripheral animal. He lives along the Mexican border, but is seen much more commonly in the Mexican portion of his range.

areas as the coatimundi and the jaguar and other cats—except that it ranges through all the southwestern states, Colorado and Utah, and up the coast through California to southern Oregon. Its tail has much plainer rings than the coatimundi.

SOME NORTHERN PERIPHERAL ANIMALS

Our northern boundary, too, with Canada, has some interesting peripheral animals, including the boundary between Alaska and Canada. One of the best known of these is the polar bear, which lives in Alaska and spreads out through the Arctic, across Canada and Greenland. Just how rare the polar bear has become is not known, but it is clearly diminishing in number. The Fish and Wildlife Service grades it as "status undetermined," and an international research program is underway to collect information. The only real threat to this animal is the "sportsman" hunter, who seems to shoot him at every opportunity, finding him by airplane and landing it to shoot him from the ground.

Four other peripherals, whose status is similarly unknown, are the fisher, the pine marten, the wolverine, and the Canada lynx. The fisher and marten are closely related animals belonging to the same family as the weasel; the fisher is the largest animal in that family. Dark brown and shaped like a weasel, it catches smaller animals for its food, especially the pine and red squirrels; it can run swiftly through treetops to catch squirrels. The fisher lives in several areas along the northern border. Greatly reduced in number, it has been rein-

The home of the polar bear changed greatly when man came with his guns. From a home with no enemies, the animal was suddenly plunged into great danger.

The wolverine is a peripheral animal of the north. This one was photographed near Juneau, Alaska.

troduced into Idaho, Oregon, Michigan, Wisconsin, and West Virginia; it is thought to be increasing in Massachusetts, New Hampshire, and northern New York. It is protected by law in Washington and Michigan.

A herd of caribou

The pine marten lives in Alaska and other northern states, and crosses the border to spread out through Canada wilderness. It also comes south in the western mountains of the United States as far as central California and northern New Mexico. It has been reintroduced in Wisconsin, Michigan, and New Hampshire, and is protected by law in Michigan.

The wolverine's home is very much in the same areas as the marten's, but includes also the high mountains of Colorado, Utah, and Idaho. It is completely protected in California, Colorado, and Washington, and is increasing in Washington.

Big cat of the north, the Canada lynx is rare along the border of the United States except that it lives in larger numbers in Alaska; and it spreads wide across Canada. It is increasing in northern Michigan, and is protected in Wisconsin and Michigan. It is most often seen in the United States when deep winter north of the border drives its food supply southward, and the lynx goes south, too, to find the rabbits and other small animals that it uses for food.

Two races of caribou are true peripherals along the Canada

border. The woodland caribou come into the northern states around the Great Lakes—a few of them. They spread from there in great numbers across Canada. Farther west, the mountain caribou—a small population of from 25 to 100—live in northern Idaho and Washington. The principal population ranges across British Columbia, in Canada, where there are many thousands.

Some of the northern peripherals are good examples of animals that have been helped, rather than harmed, by changes in their homes. In their case, the change has not been so much in the land itself, but in the conditions under which the animals lived. One change has been that man has passed laws protecting the animals, rather than hunting and trapping them ruthlessly. Another is that man has begun to recognize the importance of the animals' homes and is to some extent avoiding changes that destroy them. And still another is that man is reintroducing some of the animals into areas where they lived before man came there—where they were wiped out by the activities of man.

Nearly all of the work that man is doing to preserve the rare and threatened species among our animals has to do with maintaining and improving the animals' homes rather than destroying them.

FOR FURTHER READING

Audubon Nature Encyclopedia. 12 volumes. Curtis, 1965.

Blassingame, Wyatt. *Wonders of the Turtle World.* New York: Dodd, Mead, 1976.

Brown, Joseph E. *Wonders of Seals and Sea Lions.* New York: Dodd, Mead, 1976.

Carr, Archie. *The Everglades.* New York: Time-Life, 1973.

Chace, G. Earl. *Wonders of Prairie Dogs.* New York: Dodd, Mead, 1976.

Haines, Francis. *The Buffalo.* New York: Crowell, 1970.

Laycock, George. *The Sign of the Flying Goose, A Guide to the National Wildlife Refuges.* New York: Doubleday, 1965.

Lewin, Ted. *World Within a World: Everglades.* New York: Dodd, Mead, 1976.

The Magnificent Rockies: Crest of a Continent. Palo Alto, Ca.: American West, 1973.

Marvels and Mysteries of the World Around Us. Reader's Digest Association, 1972.

Murphy, Robert. *Wild Sanctuaries: Our National Wildlife Refuges, a Heritage Restored.* New York: Dutton, 1968.

Olson, Sigurd F. and Les Blacklock. *The Hidden Forest.* New York: Viking, 1969.

Our Amazing World of Nature. Reader's Digest Association, 1969.

Our Continent: A Natural History of North America. Washington, D.C.: National Geographic Society, 1976.

Our Living World of Nature. 10 volumes. New York: McGraw, 1966, 1967.

Rearden, Jim. *Wonders of Caribou*. New York: Dodd, Mead, 1976.

Scheffer, Victor B. *The Year of the Whale*. New York: Scribner's, 1969.

Threatened Wildlife of the United States. U.S. Department of the Interior, Fish and Wildlife Service, 1973.

Vanishing Wildlife of North America. 'Vashington, D.C.: National Geographic Society, 1974.

Wilderness U.S.A. National Geographic Society, 1973.

Wood, Dorothy. *The Bear Family*. Irvington-on-Hudson, N.Y.: Harvey, 1966.

———. *The Cat Family*. Irvington-on-Hudson, N.Y.: Harvey, 1968.

———. *The Deer Family*. Irvington-on-Hudson, N.Y.: Harvey,1969.

Wood, Frances and Dorothy Wood. *Animals in Danger: The Story of Vanishing American Wildlife*. New York: Dodd, Mead, 1968.

———. *Forests Are for People: The Heritage of Our National Forests*. New York: Dodd, Mead, 1971.

INDEX

Page numbers in **boldface** are those on which illustrations appear.